中国海洋统计年鉴

CHINA MARINE STATISTICAL YEARBOOK 2016

国家海洋局
Edited by
State Oceanic Administration,
People's Republic of China

海洋出版社
China Ocean Press

《中国海洋统计年鉴》编委会

谢　璞　　中国气象局
马云翔　　中国船舶工业集团公司
刘　悦　　中国船舶重工集团公司
刘文涛　　中国石油天然气集团公司
邓瀚深　　中国石油化工集团公司
耿　婕　　中国海洋石油总公司
刘苗夫　　中国盐业总公司
秦海岩　　中国可再生能源学会
李志军　　辽宁省海洋与渔业厅
于忠信　　河北省国土资源厅
孙连友　　天津市海洋局
宋继宝　　山东省海洋与渔业厅
张建军　　江苏省海洋与渔业局
沈依云　　上海市海洋局
童加朝　　浙江省海洋与渔业局
刘春荣　　福建省海洋与渔业厅
洪伟东　　广东省海洋与渔业厅
刘　斌　　广西壮族自治区海洋局
李向民　　海南省海洋与渔业厅
栾玉瑄　　大连市海洋与渔业局
傅道军　　青岛市海洋与渔业局
吴建义　　宁波市海洋与渔业局
王春生　　厦门市海洋与渔业局
梁俊乾　　深圳市海洋局

《中国海洋统计年鉴》编辑部

Huang Chenguang	Chinese Academy of Sciences
Xie Pu	China Meteorological Administration
Ma Yunxiang	China State Shipbuilding Corporation
Liu Yue	China Shipbuilding Industry Corporation
Liu Wentao	China National Petroleum Corporation
Deng Hanshen	China Petrochemical Group Corporation
Geng Jie	China National Offshore Oil Corporation
Liu Miaofu	China National Salt Industry Corporation
Qin Haiyan	Chinese Renewable Energy Society
Li Zhijun	Department of Ocean and Fisheries of Liaoning Province
Yu Zhongxin	Hebei Provincial Department of Land and Resources
Sun Lianyou	Tianjin Oceanic Administration
Song Jibao	Department of Ocean and Fishery of Shandong Province
Zhang Jianjun	Jiangsu Provincial Ocean and Fisheries Bureau
Shen Yiyun	Shanghai Oceanic Administration
Tong Jiachao	Zhejiang Provincial Ocean and Fisheries Bureau
Liu Chunrong	Department of Ocean and Fishery of Fujian Province
Hong Weidong	Department of Ocean and Fishery of Guangdong Province
Liu Bin	The Oceanic Administration of Guangxi
Li Xiangmin	Department of Marine and Fishery of Hainan Province
Luan Yuxuan	Dalian Ocean and Fishery Administration
Fu Daojun	Qingdao Ocean and Fishery Administration
Wu Jianyi	Ningbo Ocean and Fishery Bureau
Wang Chunsheng	Oceans and Fisheries Bureau of Xiamen
Liang Junqian	Shenzhen Oceanic Administration

Editorial Department of China Marine Statistical Yearbook

编 者 说 明

一、《中国海洋统计年鉴》2016年版是一部全面反映2015年中华人民共和国海洋经济发展和海洋管理服务情况的资料性年鉴，全书为中英文对照。

二、本年鉴的统计资料范围为人们在海洋和沿海地区开发、管理、利用海洋资源和空间，发展海洋经济的生产和活动以及沿海地区的社会经济概况。地域范围为沿海地区、沿海城市和沿海地带，其排列顺序按《沿海行政区域分类与代码》（HY/T 094-2006）的顺序排列。

三、本年鉴内容包括综合资料、海洋经济核算、主要海洋产业活动、主要海洋产业生产能力、涉海就业、海洋科学技术、海洋教育、海洋环境保护、海洋行政管理及公益服务、全国及沿海社会经济、部分世界海洋经济统计资料等十一部分。

四、本年鉴根据《海洋统计报表制度》（国统制[2016]7号）和《海洋生产总值核算制度》（国统制[2016]17号），资料主要来源于沿海省、自治区、直辖市统计局、海洋厅（局）以及20个有关涉海部、局、总公司。

五、本年鉴中除特殊说明外，所有价值量指标均为当年价，除注明年份外，其他均为2015年数据，年鉴中每部分后附有主要海洋统计指标解释，对主要海洋统计指标的含义、统计范围和统计方法做了简要说明。统计数据中的其他说明置于表的下面。有续表的资料，如有注释均置于最后一张表的下面。

六、由于第三次全国经济普查对2009-2013数据进行了修正，故本年鉴中“全国及沿海社会经济”和“海洋经济核算”中有关数据为修正后数据。

七、本年鉴表格中符号使用说明：“…”表示数据不足本表最小计算单位数；“空格”表示该项统计指标数据不详或无该项数据；“#”表示其中的主要项；其他符号如“*”或“①”等表示本表后面有注释。

八、本年鉴中由于数字精确度的原因，四舍五入后部分分项值之和与合计值有微小差异。

九、本年鉴资料国内部分未包括香港特别行政区、澳门特别行政区和台湾省数据。

十、《中国海洋统计年鉴》在编撰过程中，得到了各有关单位的大力支持，我们在此表示衷心的感谢。本年鉴中如有疏漏和不妥之处，敬请读者批评指正。

《中国海洋统计年鉴》编辑部

Introduction

Ⅰ. *China Marine Statistical Yearbook* (2016) is a data almanac which reflects in an all-round way the development of marine economy, marine management and service in the People's Republic of China in 2015, and it is a Chinese-English bilingual edition.

Ⅱ. The Yearbook's statistics cover the production and activities in the marine and coastal areas in relation to the development, management and utilization of marine resources and space, and the development of marine socioeconomy. The regions covered are the coastal regions, coastal cities and coastal zones with coastlines, which are arranged in order according to the *Coastal Administrative Areas Classification and Codes* (HY/T 094-2006).

Ⅲ. The data in the yearbook consist of 11 sections, namely, integrated data, marine economic accounting, major marine industrial activities, production capacity of major marine industries, ocean-related employment, marine science and technology, marine education, marine environmental protection, marine administration and public-good service, national and coastal socioeconomy, part of the world's marine economic statistics data.

Ⅳ. The Yearbook is based on the *Marine Statistics Report System* (Guotongzhi [2016] No. 7) and the *Ocean Gross Product Accounting System* (Guotongzhi [2016] No. 17), and its data mainly come from the statistical bureaus and the oceanic administrations of the coastal provinces, autonomous regions, and municipalities directly under the Central Government as well as the 20 ocean-related ministries, bureaus and general corporations concerned.

Ⅴ. Unless otherwise specified in the Yearbook, all the value indicators are given at the current price. Except for those noted in years, all the others are the data of 2013. Each section is attached by explanatory notes to the major marine statistical indicators, giving a brief explanation for the meaning, statistical range and statistical methods of the major marine statistical indicators. Other notes to the statistical data are listed below the tables. For the data with continued tables, annotations, if any, are put below the last table.

Ⅵ. As the Third National General Economic Survey has revised the data of 2009-2013, the related data in the "National and Coastal Scioeconomy" and "Marine Economic Accounting" of the Yearbook are the revised ones.

Ⅶ. The usage of symbols in the tables: "…" indicates the statistics smaller than the minimum calculation unit in the table; "Blank" indicates that the data of the statistical index is unknown for the time being or that there is not such data available; "#" indicates the major items of the table; Other symbols, such as "*" or "①", indicate "see footnotes below".

Ⅷ. For reasons of digital accuracy, there is small difference between the sum of values of some subterms after having been rounded off and the total values.

Ⅸ. The domestic part of the Yearbook does not include that of Hong Kong Special Administrative Region, Macau Special Administrative Region and Taiwan Province.

Ⅹ. In the course of editing *the China Marine Statistical Yearbook,* we enjoyed energetic support from the various departments concerned and we hereby extend our heartfelt thanks to them. Criticisms and comments are welcome from readers on any of the oversights and inappropriateness as the time for editing is too short.

Editorial Department of

China Marine Statistical Yearbook

目　次
CONTENTS

2015年我国海洋经济发展综述 …………………………………………………… （1）

SUMMARY OF CHINA'S MARINE ECONOMIC DEVELOPMENT IN 2015

1　综合资料

Integrated Data

1-1　沿海地区行政区划 ………………………………………………………… （27）

Administrative Division of Coastal Regions

1-2　沿海行政区划一览表 ……………………………………………………… （28）

Table of Administrative Division of Coastal Regions

1-3　海洋自然地理 ……………………………………………………………… （35）

Marine Physical Geography

1-4　海区海洋石油天然气储量 ………………………………………………… （36）

Offshore Oil and Gas Reserves in the Sea Area

1-5　沿海地区水资源情况 ……………………………………………………… （37）

Water Resources by Coastal Regions

1-6　沿海地区湿地面积 ………………………………………………………… （38）

Area of Wetlands by Coastal Regions

1-7　红树林各地类面积 ………………………………………………………… （39）

Site Classification and Area of Sharpleaf Mangrove（*Rhizophora Apiculata*）

1-8　主要沿海城市气候基本情况 ……………………………………………… （40）

Climate of Major Coastal Cities

2　海洋经济核算

Marine Economic Accounting

2-1　全国海洋生产总值 ………………………………………………………… （47）

National Gross Ocean Product

2-2　全国海洋生产总值构成 …………………………………………………… （47）

Composition of National Gross Ocean Product

2-3 海洋及相关产业增加值 …………………………………………………………（48）
Added Values of Marine and Related Industries
2-4 海洋及相关产业增加值构成 ……………………………………………………（48）
Composition of the Added Values of Marine and Related Industries
2-5 全国主要海洋产业增加值 ………………………………………………………（49）
Added Values of National Major Marine Industries
2-6 海洋渔业增加值 ……………………………………………………………………（50）
Added Value of Marine Fishery Industry
2-7 海洋油气业增加值 …………………………………………………………………（50）
Added Value of Offshore Oil and Gas Industry
2-8 海洋矿业增加值 ……………………………………………………………………（51）
Added Value of Marine Mining Industry
2-9 海洋盐业增加值 ……………………………………………………………………（51）
Added Value of Marine Salt Industry
2-10 海洋船舶工业增加值 ……………………………………………………………（52）
Added Value of Marine Shipbuilding Industry
2-11 海洋化工业增加值 ………………………………………………………………（52）
Added Value of Marine Chemical Industry
2-12 海洋生物医药业增加值 …………………………………………………………（53）
Added Value of Marine Biomedicine Industry
2-13 海洋工程建筑业增加值 …………………………………………………………（53）
Added Value of Marine Engineering Architecture
2-14 海洋电力业增加值 ………………………………………………………………（54）
Added Value of Marine Electric Power Industry
2-15 海水利用业增加值 ………………………………………………………………（54）
Added Value of Seawater Utilization Industry
2-16 海洋交通运输业增加值 …………………………………………………………（55）
Added Value of Marine Communications and Transportation Industry
2-17 滨海旅游业增加值 ………………………………………………………………（55）
Added Value of Coastal Tourism
2-18 沿海地区海洋生产总值 …………………………………………………………（56）
Gross Ocean Product by Coastal Regions

2-19 沿海地区海洋生产总值构成 …………………………………………………………（57）

Composition of Gross Ocean Product by Coastal Regions

2-20 沿海地区海洋及相关产业增加值 ……………………………………………………（58）

Added Values of Marine and Related Industries by Coastal Regions

2-21 沿海地区海洋及相关产业增加值构成 ………………………………………………（59）

Composition of the Added Values of Marine and Related Industries by Coastal Regions

3 主要海洋产业活动

Major Marine Industrial Activities

3-1 全国海洋捕捞和海水养殖产量 ……………………………………………………（67）

National Marine Catches and Mariculture Production

3-2 沿海地区海洋捕捞和海水养殖产量 ………………………………………………（68）

Marine Catches and Mariculture Production by Coastal Regions

3-3 沿海地区海洋原油产量 ……………………………………………………………（69）

Output of Offshore Crude Oil by Coastal Regions

3-4 沿海地区海洋天然气产量 …………………………………………………………（69）

Output of Offshore Natural Gas by Coastal Regions

3-5 海洋原油出口量及创汇额 …………………………………………………………（70）

Export Volume of and Foreign-Exchange Earnings from Offshore Crude Oil by Coastal Regions

3-6 海洋原油产量、出口量占全国原油产量、出口量比重 ……………………………（70）

Proportion of Offshore Crude Oil Production and Export Volume in the National Total

3-7 沿海地区海洋矿业产量 ……………………………………………………………（71）

Output of Marine Mining Industry by Coastal Regions

3-8 沿海地区海盐产量 …………………………………………………………………（72）

Sea Salt Production by Coastal Regions

3-9 沿海地区海洋化工产品产量 ………………………………………………………（73）

Output of Marine Chemical Products by Coastal Regions

3-10 沿海地区海洋生物医药产品产量 …………………………………………………（74）

Production of the Marine Biomedicine Industry by Coastal Regions

3-11 沿海地区海洋修造船完工量 ………………………………………………………（76）

Production of the Marine Shipbuilding Industry by Coastal Regions

3-12 沿海地区海洋货物运输量和周转量 ………………………………………………（77）

Volume of Maritime Goods Transported and Turnover by Coastal Regions

3-13 沿海地区海洋旅客运输量和周转量 ………………………………………………………… (78)
Volume of Maritime Passenger Traffic and Turnover by Coastal Regions
3-14 沿海港口客货吞吐量 ………………………………………………………………………… (79)
Volume of Passenger and Freight Handled at Coastal Seaports
3-15 沿海地区水路国际标准集装箱运量 ………………………………………………………… (80)
Volume of International Standardized Containers Traffic by Coastal Regions
3-16 沿海港口国际标准集装箱吞吐量 …………………………………………………………… (81)
International Standardized Containers Handled at Coastal Seaports
3-17 沿海城市国内旅游人数 ……………………………………………………………………… (82)
Domestic Visitors by Coastal Cities
3-18 主要沿海城市接待入境游客人数 …………………………………………………………… (84)
Number of Inbound Tourists Received by Major Coastal Cities
3-19 主要沿海城市接待入境游客情况 …………………………………………………………… (85)
Breakdown of Inbound Tourists Received by Major Coastal Cities

4 主要海洋产业生产能力

Production Capacity of Major Marine Industries

4-1 沿海渔港情况 …………………………………………………………………………………… (93)
Coastal Fishing Ports
4-2 沿海地区海水养殖面积 ………………………………………………………………………… (94)
Mariculture Area by Coastal Regions
4-3 海洋油气勘探情况 ……………………………………………………………………………… (95)
Work Volume of Offshore Oil and Gas
4-4 海洋油气生产井情况 …………………………………………………………………………… (95)
Survey of Offshore Oil and Gas Production Wells
4-5 沿海地区盐田面积和海盐生产能力 …………………………………………………………… (96)
Salt Pan Area and Sea Salt Production Capacity by Coastal Regions
4-6 海上风电开工项目 ……………………………………………………………………………… (97)
Offshore Wind Power Generation Projects
4-7 主要潮汐电站分布情况 ………………………………………………………………………… (98)
Distribution of Major Tidal Power Stations
4-8 主要海上活动船舶 ……………………………………………………………………………… (99)
Major Vessels Operating on the Sea

4-9 沿海规模以上港口生产用码头泊位 …………………………………………… (100)
Berths for Productive Use at the Coastal Seaports Above Designed Size
4-10 沿海地区星级饭店基本情况 …………………………………………………… (102)
Star Grade Hotels and Occupancies by Coastal Regions
4-11 沿海地区旅行社数 …………………………………………………………… (103)
Number of Travel Agencies by Coastal Regions
5 涉海就业
Ocean-Related Employment
5-1 全国涉海就业人员情况 ……………………………………………………… (109)
National Ocean-Related Employed Personnel by Regions
5-2 全国主要海洋产业就业人员情况 …………………………………………… (110)
Employed Personnel in the Major Marine Industries throughout the Country
6 海洋科学技术
Marine Science and Technology
6-1 分行业海洋科研机构及人员情况 …………………………………………… (113)
Marine Scientific Research Institutions and Personnel by Industry
6-2 分行业海洋科研机构科技活动人员学历构成 ……………………………… (115)
Educational Background Composition of the Personnel Engaged in Scientific and Technological Activities in the Marine Scientific Research Institutions by Industry
6-3 分行业海洋科研机构科技活动人员职称构成 ……………………………… (117)
Technical Title Composition of Personnel Engaged in Scientific and Technological Activities in the Marine Scientific Research Institutions by Industry
6-4 分行业海洋科研机构经费收入 ……………………………………………… (119)
Routine Fund Receipts of the Marine Scientific Research Institutions by Industry
6-5 分行业海洋科研机构科技课题情况 ………………………………………… (121)
Marine Science and Technology Research Projects of the Research Institutions by Industry
6-6 分行业海洋科研机构科技论著情况 ………………………………………… (123)
Marine Scientific and Technological Works of the Research Institutions by Industry
6-7 分行业科研机构科技专利情况 ……………………………………………… (125)
Marine Scientific and Technological Patents of the Research Institutions by Industry
6-8 分行业海洋科研机构R & D情况 …………………………………………… (127)
R & D in the Marine Scientific Research Institutions by Professions

6-9 分地区海洋科研机构及人员情况 ………………………………………………………… (129)
Marine Scientific Research Institutions and Personnel by Regions
6-10 分地区海洋科研机构科技活动人员学历构成 …………………………………… (130)
Educational Background Composition of the Personnel Engaged in Scientific and Technological Activities in the Marine Scientific Research Institutions by Regions
6-11 分地区海洋科研机构科技活动人员职称构成 …………………………………… (131)
Technical Title Composition of Personel Engaged in the Marine Scientific Research Institutions by Regions
6-12 分地区海洋科研机构经费收入 ………………………………………………………… (132)
Routine Fund Receipts of the Marine Scientific Research Institutions by Regions
6-13 分地区海洋科研机构科技课题情况 …………………………………………………… (133)
Marine Science and Technology Research Projects of the Research Institutions by Regions
6-14 分地区海洋科研机构科技论著情况 …………………………………………………… (134)
Marine Scientific and Technological Works of the Research Institutions by Regions
6-15 分地区海洋科研机构科技专利情况 …………………………………………………… (135)
Marine Scientific and Technological Patents of the Research Institutions by Regions
6-16 分地区海洋科研机构R & D情况 ……………………………………………………… (136)
R & D in the Marine Scientific Research Institutions by Regions
7 海 洋 教 育
Marine Education
7-1 全国各海洋专业博士研究生情况 ……………………………………………………… (143)
Doctoral Students from Marine Specialities
7-2 全国各海洋专业硕士研究生情况 ……………………………………………………… (145)
Postgraduate Students from Marine Specialities
7-3 全国普通高等教育各海洋专业本科学生情况 ………………………………………… (147)
Undergraduates from Marine Specialities in the National Ordinary Higher Education
7-4 全国普通高等教育各海洋专业专科学生情况 ………………………………………… (148)
Students from Marine Specialities of the Colleges for Professional Training in the National Ordinary Higher Education
7-5 全国成人高等教育各海洋专业本科学生情况 ………………………………………… (150)
Undergraduates from Marine Specialities in the National Adult Higher Education
7-6 全国成人高等教育各海洋专业专科学生情况 ………………………………………… (151)
Students from Marine Specialities of the Colleges for Professional Training in the National Adult Higher Education

7-7 全国中等职业教育各海洋专业学生情况 …………………………………… （153）
Students from Marine Specialities in the National Secondary Vocational Education
7-8 分地区各海洋专业博士研究生情况 ……………………………………… （155）
Doctoral Students in Marine Specialities by Regions
7-9 分地区各海洋专业硕士研究生情况 ……………………………………… （156）
Postgraduate Students in Marine Specialities by Regions
7-10 分地区普通高等教育各海洋专业本科学生情况 ………………………… （157）
Undergraduates in the Marine Specialities of Ordinary Higher Education by Regions
7-11 分地区普通高等教育各海洋专业专科学生情况 ………………………… （158）
Students of the Marine Specialities Colleges for Professional Training in the Ordinary Higher Education by Regions
7-12 分地区成人高等教育各海洋专业本科学生情况 ………………………… （159）
Students from Marine Specialities in the Adult Higher Education by Regions
7-13 分地区成人高等教育各海洋专业专科学生情况 ………………………… （160）
Students from Marine Specialities of the Colleges for Professional Training in the Adult Higher Education by Regions
7-14 分地区中等职业教育各海洋专业学生情况 ……………………………… （161）
Students from Marine Specialities in the Secondary Vocational Education by Regions
7-15 分地区开设海洋专业高等学校教职工数 ………………………………… （162）
Number of Teaching and Administrative Staff in the Universities and Colleges Offering Marine Specialities by Regions

8 海洋环境保护

Marine Environmental Protection

8-1 海区海水水质评价结果 ……………………………………………………… （167）
Seawater Quality Assessment Results
8-2 海区废弃物海洋倾倒情况 …………………………………………………… （169）
Ocean Dumping of Wastes by Sea Area
8-3 海区海洋石油勘探开发污染物排放入海情况 ……………………………… （170）
Discharge of Pollutants into the Sea from Offshore Oil Exploration and Exploitation
8-4 沿海地区工业废水排放及处理情况 ………………………………………… （171）
Discharge and Treatment of Industrial Waste Water by Coastal Regions

8-5 沿海城市工业废水排放及处理情况 …………………………………………… (172)
Discharge and Treatment of Industrial Waste Water by Coastal Cities
8-6 沿海地带工业废水排放及处理情况 …………………………………………… (175)
Discharge and Treatment of Industrial Waste Water by Coastal Counties
8-7 沿海地区一般工业固体废物倾倒丢弃、处理及综合利用情况 ……………… (178)
Discharge, Treatment and Multipurpose Utilization of Common Industrial Solid Wastes by Coastal Regions
8-8 沿海城市一般工业固体废物倾倒丢弃、处理及综合利用情况 ……………… (179)
Discharge, Treatment and Multipurposed Utilization of Common Industrial Solid Wastes by Coastal Cities
8-9 沿海地带一般工业固体废物倾倒丢弃、处理及综合利用情况 ……………… (182)
Discharge, Treatment and Multipurposed Utilization of Common Industrial Solid Wastes by Coastal Counties
8-10 主要沿海城市工业废气排放及处理情况 ……………………………………… (185)
Emission and Treatment of Industrial Waste Gas in Major Coastal Cities
8-11 沿海地区污染治理项目情况 ……………………………………………………… (191)
Pollution Treatment Projects in Coastal Regions
8-12 沿海城市污染治理项目情况 ……………………………………………………… (192)
Pollution Treatment Projects in Coastal Cities
8-13 沿海地带污染治理项目情况 ……………………………………………………… (195)
Pollution Treatment Projects in Coastal Counties
8-14 沿海区域海洋类型自然保护区建设情况 ………………………………………… (198)
Construction of Marine-Type Nature Reserves in Coastal Regions
8-15 沿海地区海洋类型自然保护区建设情况 ………………………………………… (199)
Construction of Marine-Type Nature Reserves
8-16 全国海洋生态监控区基本情况 …………………………………………………… (200)
Basic Condition of the Marine Ecological Monitoring Areas Throughout the Country
8-17 沿海地区风暴潮灾害情况 ………………………………………………………… (202)
Survey of Storm Surges by Coastal Regions
8-18 沿海地区赤潮灾害情况 …………………………………………………………… (204)
Survey of Red Tide by Coastal Regions

9 海洋行政管理及公益服务

Marine Administration and Public-Good Service

9-1 海域使用管理情况 …………………………………………………… (209)

Sea Area Use Management

9-2 依法批准无居民海岛开发利用情况 ………………………………… (210)

Development and Utilization of Uninhabited Islands Legally Approved

9-3 海洋倾废管理情况 …………………………………………………… (211)

Management on the Ocean Dumping of Wastes

9-4 海洋行政执法情况 …………………………………………………… (212)

Marine Administrative Enforcement

9-5 沿海地区海滨观测台站分布概况 …………………………………… (213)

Distribution of Coastal Observation Stations by Coastal Regions

9-6 海洋预报服务概况 …………………………………………………… (214)

Marine Forecast Service

9-7 海洋环境观测情况 …………………………………………………… (216)

Condition of Marine Environmental Observation

9-8 海洋调查概况 ………………………………………………………… (216)

Marine Survey Statistics

9-9 涉外海洋科学研究审批情况 ………………………………………… (217)

Examination and Approval of Foreign-Related Marine Scientific Research

9-10 海洋档案及利用情况 ……………………………………………… (217)

Marine File and Its Use

9-11 卫星遥感接收应用情况 …………………………………………… (218)

Remote-Sensing Receiving and Utilization

9-12 海洋标准化监督管理情况 ………………………………………… (219)

Supervision and Management of Marine Standardization

10 全国及沿海社会经济

National and Coastal Socioeconomy

10-1 国内生产总值 ……………………………………………………… (225)

Gross Domestic Product

10-2 国内生产总值增长速度 …………………………………………… (225)

Growth Rate of Gross Domestic Product

10-3 沿海地区生产总值 …………………………………………………………………… (226)
Gross Regional Product of Coastal Regions
10-4 沿海地区生产总值增长速度 ……………………………………………………… (227)
Growth Rate of Gross Regional Product of Coastal Regions
10-5 沿海城市生产总值(2014年) ……………………………………………………… (228)
Gross Regional Product of Coastal Cities, 2014
10-6 县级单位主要统计指标 …………………………………………………………… (230)
Main Indicators of Regions at County Level
10-7 沿海地区财政收支 ………………………………………………………………… (240)
Financial Revenue and Expenditure by Coastal Regions
10-8 沿海地区教育基本情况 …………………………………………………………… (241)
Basic Conditions of Education by Coastal Regions
10-9 沿海地区卫生基本情况 …………………………………………………………… (242)
Basic Conditions of Public Health by Coastal Regions
10-10 分地区电力消费量 ……………………………………………………………… (243)
Electricity Consumption by Region
10-11 沿海地区用水情况 ……………………………………………………………… (244)
Water Use of Coastal Regions
10-12 沿海地区全社会固定资产投资 ………………………………………………… (245)
Total Investment in Fixed Assets by the Whole Society of Coastal Regions
10-13 沿海地区按经营单位所在地分货物进出口总额 ……………………………… (246)
Total Value of Imports and Exports by Location of Importers/Exporters of Coastal Regions
10-14 沿海地区城镇居民平均每人全年家庭收入和消费性支出 …………………… (247)
Per Capita Annual Income and Consumption Expenditure of Urban Households by Coastal Regions
10-15 沿海地区农村居民可支配收入和生活消费支出 ……………………………… (248)
Per Capita Disposable Income and Consumption Expenditure of Rural Residents by Coastal Regions
10-16 沿海地区年末人口数 …………………………………………………………… (249)
Year-end Population by Coastal Regions
10-17 沿海城市人口情况(2014年) …………………………………………………… (250)
Population of Coastal Cities, 2014

10-18 沿海县人口情况 ………………………………………………………… (253)
Population of Coastal Counties
10-19 沿海地区城镇单位就业人员情况 ……………………………………… (258)
Employed Persons in Urban Units by Coastal Regions
10-20 沿海城市就业人员情况(2014年) ……………………………………… (259)
Number of Employed Persons by Coastal Cities, 2014
11 世界海洋经济统计资料(部分)
World's Marine Economic Statistics Data (Part)
11-1 世界海洋面积(2010年) ……………………………………………… (271)
World Ocean Area, 2010
11-2 主要沿海国家(地区)海岸线长度 …………………………………… (272)
Length of Coastline of Major Coastal Countries (Areas)
11-3 世界主要沿海国家(地区)国土面积和人口 ………………………… (273)
Surface Area and Population of Major Coastal Countries (Areas)
11-4 主要沿海国家(地区)国内生产总值 ………………………………… (274)
Gross Domestic Product of Major Coastal Countries (Areas)
11-5 主要沿海国家(地区)国内生产总值产业构成 ……………………… (275)
Industrial Composition of GDP of Major Coastal Countries (Areas) by Industry
11-6 主要沿海国家(地区)就业人数 ……………………………………… (277)
Employment in the Major Coastal Countries (Areas)
11-7 主要沿海国家(地区)鱼类产量(2014年) …………………………… (277)
Fish Yields of Major Coastal Countries (Areas), 2014
11-8 国家保护区面积和鱼类濒危物种 …………………………………… (278)
Area of National Nature Reserves and Endangered Species of Fish
11-9 主要沿海国家(地区)风力发电量 …………………………………… (279)
Wind-Power Capacities of Major Coastal Countries (Areas)
11-10 主要沿海国家(地区)捕捞产量 ……………………………………… (280)
Fishing Yields in the Major Coastal Countries (Areas)
11-11 主要沿海国家(地区)水产养殖产量 ………………………………… (281)
Aquaculture Production in the Major Coastal Countries (Areas)
11-12 主要沿海国家(地区)石油主要指标 ………………………………… (282)
Main Oil Indicators of Major Coastal Countries (Areas)

11-13 万美元国内生产总值能耗
Energy Use per Ten Thousand USD of GDP …………………………………… (283)
11-14 主要沿海国家(地区)国际海运装货量和卸货量 ………………………… (284)
International Maritime Freight Loaded and Unloaded in the Major Coastal Countries (Areas)
11-15 世界主要外贸货物海运量及构成(2014年) ……………………………… (285)
World Major Maritime Freight Traffic in Foreign Trade, 2014
11-16 主要沿海国家(地区)国际旅游人数 ……………………………………… (286)
Number of International Tourists of Major Coastal Countries (Areas)
11-17 主要沿海国家(地区)国际旅游收入 ……………………………………… (288)
International Tourism Receipts of Major Coastal Countries (Areas)
11-18 集装箱吞吐量居世界前20位的港口 ……………………………………… (290)
World Top 20 Seaports in Terms of the Number of Containers Handled
11-19 港口货物吞吐量居世界前20位的港口(2014年) ………………………… (291)
World Top 20 Seaports in Terms of the Cargo Handled, 2014
11-20 海上商船拥有量居世界前20位的国家或地区(2014年) ………………… (292)
World Top 20 Countries or Areas in Terms of the Number of Maritime Merchant Ships Owned, 2014
11-21 集装箱船拥有量居世界前20位的国家或地区(2014年) ………………… (293)
World Top 20 Countries or Areas in Terms of the Number of Container Ships Owned, 2014
11-22 海上油轮拥有量居世界前20位的国家或地区(2014年) ………………… (294)
Top 20 Countries or Areas in the World by the Possession of Ocean-Going Oil Tankers, 2014
11-23 海上散货船拥有量居世界前20位的国家或地区(2014年) ……………… (295)
Top 20 Countries or Areas in the World by the Possession of Ocean Bulk Carriers, 2014

2015年我国海洋经济发展综述

2015年，沿海地区各级人民政府、海洋行政主管部门认真贯彻落实党中央、国务院建设海洋强国和21世纪海上丝绸之路、壮大海洋经济的战略部署，全面推进《全国海洋经济发展“十二五”规划》实施，海洋产业结构调整步伐加快，海洋经济在新常态下总体保持平稳增长态势。

一、全国海洋经济发展概况

据初步核算，2015年全国海洋生产总值65 534亿元，比上年增长7.0%（除特别注明外，增长率均按可比价计算），海洋生产总值占国内生产总值的9.5%，占沿海地区生产总值的16.8%。全国涉海就业人员3 588.5万人，比上年增加34.8万人。

二、主要海洋产业发展情况

2015年，主要海洋产业实现增加值26 839亿元，比上年增长6.1%，占海洋生产总值的41.0%，滨海旅游业和海洋交通运输业仍占主导地位。

海洋第一产业　2015年，海洋渔业保持平稳发展态势，海洋水产品产量稳步增长，海水养殖及远洋渔业生产能力持续提高。海洋水产品产量为3 409.6万吨，比上年增长3.4%。其中，海水养殖产量继续增加，达到1 875.6万吨，比上年增长3.5%；海洋捕捞产量也有所提高，为1 314.8万吨，比上年增长2.7%；由于远洋捕捞能力的加强，产量快速增加，达219.2万吨，比上年增长8.1%。海水养殖面积有所上升，为231.8万公顷，比上年增长

0.5%。远洋渔船数量达到 2 512 艘，比上年增加 2.1%；总功率达到 215.7 万千瓦，比上年增长 6.5%。

海洋第二产业 2015 年，我国海洋油气产量依然保持增长，但受国际油价持续走低等因素影响，全年实现增加值 981.9 亿元，比上年增长 2.4%，海洋原油产量 5 416 万吨，比上年增长 17.4%，海洋天然气产量 147.2 亿立方米，比上年增长 11.3%。海洋矿业保持稳定增长，全年实现增加值 63.9 亿元，比上年增长 7.6%。海洋盐业产能过剩情况严重，消费需求不足，行业效益出现明显下滑，全年实现增加值 61.0 亿元，比上年减少 25.4%。海洋化工业较快增长，全年实现增加值 964.2 亿元，比上年增长 12.4%。海洋生物医药业快速增长，全年实现增加值 295.7 亿元，比上年增长 12.4%。海洋电力业继续较快增长，海上风电场建设有序推进，全年实现增加值 120.1 亿元，比上年增长 13.0%。海水利用业保持平稳的增长态势，发展环境持续向好，全年实现增加值 13.7 亿元，比上年增长 9.5%。海洋船舶工业加速淘汰落后产能，转型升级成效明显，但仍面临较为严峻的形势，全年实现增加值 1 445.7 亿元，比上年增长 1.4%。海洋工程建筑业快速发展，全年实现增加值 2 073.5 亿元，比上年增长 13.1%。

海洋第三产业 2015 年，全国沿海港口生产形势总体良好，但受全球经济增长放缓、供需失衡问题持续存在的影响，国内航运市场仍在低迷与复苏之间徘徊，全年海洋交通运输业实现增加值 5 641.1 亿元，比上年增长 7.6%。沿海港口货物吞吐量 814 728 万吨，比上年增长 1.4%，国际标准集装箱吞吐量 18 907 亿标准箱，比上年增长 4.0%。滨海旅游产业规模继续扩大，沿海地区不断拓展海洋旅游新项目，提升海洋旅游服务质量和水平，海洋邮轮游艇旅游发展成为海洋旅游消费新热点。主要沿海城市接待入境旅游者人数 4 064.1 万人次，比上年增加 177.5 万人次，其中港澳台入境游客占 56.9%，是主要的客源市场；滨海旅游业全年实现增加值 10 880.6 亿元，比上年增长 11.5%。

三、区域海洋经济发展情况

2015 年，环渤海、长江三角洲和珠江三角洲三大经济区海洋经济继续保持平稳增长态势，但增速均有所放缓，三大经济区占全国海洋生产总值的比重分别为 35.1%、28.8%和 22.0%。

环渤海经济区海洋生产总值 23 002.7 亿元，比上年增长 3.2%（现价），占地区生产总值比重 17.4%。海洋产业增加值 13 713.3 亿元，海洋相关产业增加值为 9 289.4 亿元。海洋渔业、海洋油气业、海洋交通运输业、滨海旅游业是该地区海洋经济的支柱产业，增加值合计达到 8 634.3 亿元，占该地区主要海洋产业增加值的 82.9%。海洋工程建筑业、滨海旅游业和海洋生物医药业实现快速增长，分别比上年增长 29.7%、7.5%和 12.9%（现价）。

长江三角洲经济区海洋生产总值 18 878.0 亿元，比上年增长 9.3%（现价），占地区生产总值比重 13.7%。海洋产业增加值 11 248.0 亿元，海洋相关产业增加值 7 630.0 亿元。滨海旅游业、海洋交通运输业、海洋船舶工业和海洋渔业四个产业位居各产业前列，其增加值之和占该地区主要海洋产业增加值的 93.0%，其中滨海旅游业占该地区主要海洋产业增加值的 43.5%，产值贡献位居第一。海洋生物医药业、海洋电力业增速较快，分别比上年增长 16.1%和 18.7%（现价）。

珠江三角洲经济区海洋生产总值 14 443.1 亿元，比上年增长 9.2%（现价），占地区生产总值比重达 19.8%。海洋产业增加值为 9 085.8 亿元，海洋相关产业增加值为 5 357.4 亿元。滨海旅游业、海洋交通运输业、海洋化工业和海洋工程建筑业四个产业位居前列，其增加值之和占该地区主要海洋产业增加值的 82.4%。滨海旅游业、海洋工程建筑业和海洋生物医药业增长较快，分别比上年增长 13.2%、10.2%和 10.0%（现价）。

四、科技教育

2015 年，海洋科研教育事业继续保持稳步发展，统计的海洋科研机构共 192 个，从业人员 42 331 人，比上年增长 4.4%；海洋科研机构承担海洋科技课题 18 810 项，比上年增长 6.3%；发表海洋科技论文 17 257 篇，出版海洋科技著作 353 种；专利授权数 5 622 件，其中发明专利 4 111 件，比上年增长 55.3%。开设海洋专业的高等院校达 415 个，比上年增加 7 个，专任教师数 322 942 人，比上年增长 5.3%。高等教育和中等职业教育海洋专业毕业生数 75 011 人，比上年下降 9.2%；招生人数和在校人数分别为 67 651 人和 219 328 人，分别比上年下降 5.7%和 5.9%。

五、海洋环境保护与防灾减灾

2015 年，我国海洋环境质量总体基本稳定，污染主要集中在近岸局部海域，我国近岸局部海域污染依然严重，冬季、春季、夏季和秋季劣于第四类海水水质标准的海域面积分别为 67 150、51 740、40 020 和 63 230 平方千米。实施监测的河口、海湾、滩涂湿地、珊瑚礁、红树林和海草床等海洋生态系统中，处于健康、亚健康和不健康状态的海洋生态系统分别占 14%、76%和 10%。我国海洋灾情总体偏轻，海洋灾害共造成直接经济损失 72.7 亿元，死亡（含失踪）30 人。其中，造成直接经济损失最严重的是风暴潮灾害，占总直接经济损失的 99.8%；赤潮发生次数与累计面积均为近五年最低值。

六、海洋行政管理

2015 年，行政管理工作扎实推进，各项海洋工作取得明显成效。全年颁发海域使用权证书 3 606 本，比上年下降 28.0%；确权海域面积 253 613 公顷，比上年减少 32.2%；全年共签发疏浚物海洋倾倒许可证 100 份，实施各项海洋执法检查共 27 686 次，发现违法行为 75 起；提供海洋数值预报服务共 35 412 次，开展海洋调查项目 785 个，获得数据共 32.4 万个，各项海洋观测获得数据量共 36 370.1 万个，全年接收存档卫星遥感数据量共计 80 884.5GB；本年接收纸质档案 152 938 卷（册），电子档案 37 108GB；共有 22 项国家标准和 72 项行业标准通过立项审查，出版国家标准 7 项，行业标准 21 项。

Summary of China´s Marine Economic Development in 2015

In 2015, the people's governments of the coastal areas and the departments of Marine administration at all levels conscientiously implement the strategy and development of the Party Central Committee, the State Council to build a marine power of the sea and marine silk road of the 21st century, and expand the marine economy, comprhensively promote *National Marine Economy Development five - year Plan*, accelerate the adjustment of the structure of the marine industry, and the marine economy has maintained steady growth under the new normal.

Ⅰ. Survey of the National Marine Economic Development

In 2015, according to the preliminary accounting, the national Gross Ocean Product amounted to 6 553.4 billion yuan, 7.0% up from the previous year (Unless specified otherwise, the growth rate is calculated at the comparable price), accounting for 9.5% of the national GDP and 16.8% of the Gross Coastal Product. The number of people employed by the marine-related sectors throughout the country reached 35.885 million, 0.348 million more than that in the previous year.

Ⅱ. Development of Major Marine Industries

The major marine industries effected an added value of 2 683.9 billion yuan, registering an increase of 6.1% as against that last year, accounting for 41.0% of the Gross Ocean Product, and coastal tourism and marine communication and transport still occupied the leading position.

Primary Marine Industry

In 2015, marine fishery kept a posture of sustained and stable development, the output of marine aquatic products steadily increased and the capacity of mariculture and ocean-going fishery continued to be strengthened. The yield of marine aquatic products amounted to 34.096 million tons, 3.4% up from the previous year, among which, the output of mariculture continued to grow, reaching 18.756 million tons, 3.5% up from that last year; the yield from marine fishing increased to some extent, amounting to 13.148 million tons, 2.7% up from the previous year. Thanks to the enhancement of the ocean-going fishing capacity, the output witnessed a rapid growth, reaching 2.192 million tons, 8.1% up from the previous year; the area of mariculture somewhat increased, amounting to 2.318 million hm^2, increasing by 0.5% as compared with that last year; the number of ocean-going fishing vessels reached 2 512, 2.1% up from the previous year; and the total power amounted to 2.157 million kW, 6.5% up from that last year.

Secondary Marine Industry

In 2015, China's output of offshore oil and gas still kept growing, but affected by such factors as continued dropping of international oil price, it effected an added value of 98.2 billion yuan, 2.4% up from the previous year, the output of marine crude oil was 54.16 million tons, 17.4% up from the previous year and that of offshore natural gas 14.72 billion m^3, growing by 11.3% as against that in 2014. The marine mining industry kept growing steadily, effecting a full-year added value of 6.4 billion yuan, 7.6% up from the previous year. The marine salt industry has serious overcapacity, low consumption demand and obvious decline in industry benefits, realizing a full-year added value of 4.1 billion yuan, 25.4% down from the previous year; the marine chemical industry grew rapidly, accomplishing an added value of 96.4 billion yuan for the whole year, 12.4% up from the previous year; the marine biomedicine industry grew rapidly, effecting a full-year added value of 29.6 billion yuan, growing by 14.0% as compared with that in 2014; the marine electric power industry developed steadily and the construction offshore wind farms was carried forward steadily, realizing a full-year added value of 12.0 billion yuan, 13.0% up from the previous year; the seawater utilization industry maintained a steadily growing posture and the development environment continued to turn for the better, effecting an added value of 1.4 billion yuan throughout the year, 9.5% up from the previous year. The marine shipbuilding industry has been speeding up the elimination of backward production capacity, with obvious achivements in restructing and upgrading, but still faces a severe situation. The anuual added value of 144.5 billion yuan, 1.4% up from the previous year; and the marine engineering construction industry development rapidly, realizing an added value of 207.4 billion yuan for the whole year, 13.1% up from 2014.

Tertiary Marine Industry

In 2015, the production situation of the coastal ports in China was good on the whole, but affected by the slowing down of the global economic growth and the persistent existence of the problem of imbalance between supply and demand, the domestic shipping market still hovered between recession and recovery and the marine communications and transport accomplished a value added of 554.1 billion yuan for the whole year, 5.6% up from the previous year. The cargo handling capacity of coastal ports was 8 147.28 million tons, registering a growth rate of 1.4% over year, and the handling capacity of international standardized containers 1 890.7 billion standard boxes, 4.0% up from 2014. The scale of coastal tourism continued to expand, the coastal regions constantly expanded new projects of marine tourism and upgraded the service quality and level of marine tourism, and the development of marine passenger liners and yachting tourism became a new hot spot for marine tourist consumption. The number of the inbound tourists received by the major coastal cities amounted to 40.641 million person-times, 1.78 million person-times more than that in 2014, of which, the inbound tourists from Hongkong, Macao and Taiwan occupied 56.9%, which constituted the major market of source of tourists; the coastal tourism industry accomplished a full-year added value of 1 087.41 billion yuan, 11.4% up from 2014.

Ⅲ. Development of Regional Marine Economy

In 2015, the marine economy in the three major economic zones of the Round-the-Bohai, Changjiang River Delta and Zhujiang River Delta continued to maintain a posture of steady growth, but their growth rates all slowed down to some extent and the proportions of the three major economic zones in the national Gross Ocean Product were 35.1%, 28.8% and 22.0% respectively.

The Gross Ocean Product of the Round-the-Bohai economic zone was 2 300.27 billion yuan, registering a growth rate of 3.2% over the previous year (at the current price), accounting for 17.4% of the Gross Regional Product, The value added of marine industries was 1371.3 billion yuan and that of the marine-related industries 928.94 billion yuan. Marine fishery, offshore oil and gas industry, marine communications and transport, and coastal tourism were the pillar industries in the zone, their added values adding up to 863.43 billion yuan and accounting for 82.9% of the added value of the region's marine industries. The marine engineering construction, coastal tourism and marine biomedicine industries accomplished a rapid growth, 29.7%,

7.5% and 12.9% up from the previous year respectively (at the current price).

The Gross Ocean Product of the Changjiang River Delta Economic Zone amounted to 1 878.00 billion yuan, 9.3% up to from the previous year (at the current price) and accounting for 13.7% of the Gross Regional Product. The value added of the marine industries was 1 124.80 billion yuan and that of the marine-related industries 763.00 billion yuan. The four industries of coastal tourism, marine communications and transport, marine shipbuilding and marine fishery ranked in the forefront of all industries, their added values adding up to 92.9% of the added value of the regional's major marine industries, of which, that of the coastal tourism accounted for 43.5% of the added value of the region's major marine industries and ranked first in the product contribution. The marine biomedicine and marine electric power industries grew rapidly, 16.1% and 18.7% respectively up from the previous year (at the current price).

The Gross Ocean Product of the Zhujiang River Delta Economic Zone was 1 444.31 billion yuan, 9.2% up from the previous year (at the current price) and accounting for 19.8% of the Gross Regional Product. The added value of marine industries was 908.58 billion yuan and that of the marine-related industries 535.7 billion yuan. The four industries of coastal tourism, marine communications and transport, marine chemical industry and marine engineering construction were among the front-runners, their added values adding up to 82.4% of that of the region's major marine industries. Coastal tourism, marine engineering construction and marine biomedicine grew rapidly, 13.2%, 10.2%, and 10.0% up from the previous year, respectively(at the current price).

Ⅳ. Marine Scientific Research and Education

In 2015, the marine scientific research and education continued to develop steadily. The statistics number of marine scientific research institutions totalled 192 with 42 331 employees, increasing by 4.4% as compared with the previous year; the number of marine scientific and technological projects undertaken by these institutions was 18 801, 6.3% up from last year; 17 257 marine scientific and technological papers and 353 kinds of marine scientific and technological works were published; the number of patents authorized was 5 622, of which 4 111 were the ones for discovery, 55.3% up from the previous year; the number of the institutions of higher learning which offer marine specialities amounted to 415, 7 more than last year, and the number of full-time teachers in this regard was 322 942, 5.3% up from the previous year; the number of graduates from the marine speciality of higher learning and secondary vocational education was 75 011, 9.2% down from last year; and the number of students enrolled and that in school were 67 651 and 219 328, 5.7% and 5.9% down from the previous year separately.

Ⅴ. Marine Environmental Protection, and Disaster Prevention and Mitigation

In 2015, the state of China's marine environmental quality was basically stable on the whole. Pollution was mainly concentrated in the local nearshore areas, and was still serious there. The sea areas with the seawater quality inferior to Class 4 amounted to 67 150, 51 740, 40 020 and 63 230 km^2 respectively in spring, summer, autumn and winter. Among the marine ecosystems that had been monitored such as estuaries, tidal flat wetland, coral reef, mangrove and sea grass bed etc., those in the healthy, sub-healthy or unhealthy states accounted for 14%, 76% and 10% respectively. The effects of marine disasters in China in 2015 were on the whole on the light side, the direct economic loss caused by marine disasters reached 7.27 billion yuan and the death (including missing) of 30 persons was resulted therefrom. Among others, the storm surge disaster was the one which caused the most serious direct economic loss, which accounted for 99.8% of the total direct economic loss; the occurring frequency and the total area of red tides were both the lowest in the past five years.

Ⅵ. Marine Administration

In 2015, marine administration was soundly carried forward and various marine effort obtained obvious results. A total of 3 606 certificates for the sea area use right were issued for the whole year, 28.0%

down from 2014; the sea area with the ownership of patent rights reached 253 613 hm^2, 32.2% down from the previous year, a total of 100 permits for oceanic dumping of dredged material were signed and issued throughout the year, and marine inspections for law enforcement were carried out on 27 686 occasions, during which 75 cases of unlawful practice were discovered; the marine numerical forecast service was provided for 35 412 times, 785 marine survey projects were carried out, acquiring 0.324 million data and the data quantity from various marine observations totalled 363.701 million and the quantity of satellite remote-sensing data received and placed on file for the whole year was 80 884.5GB; the paper archives received this year totalled 152 938 volumes (copies), and the electronic archives 37 108 GB; and a total of 22 items of national standard and 72 items of professional standard had been authorized and examined, and 7 items of national standard and 21 items of professional items published.

图 1　全国海洋生产总值及三次产业构成

China's Gross Ocean Product and Three Industries Composition

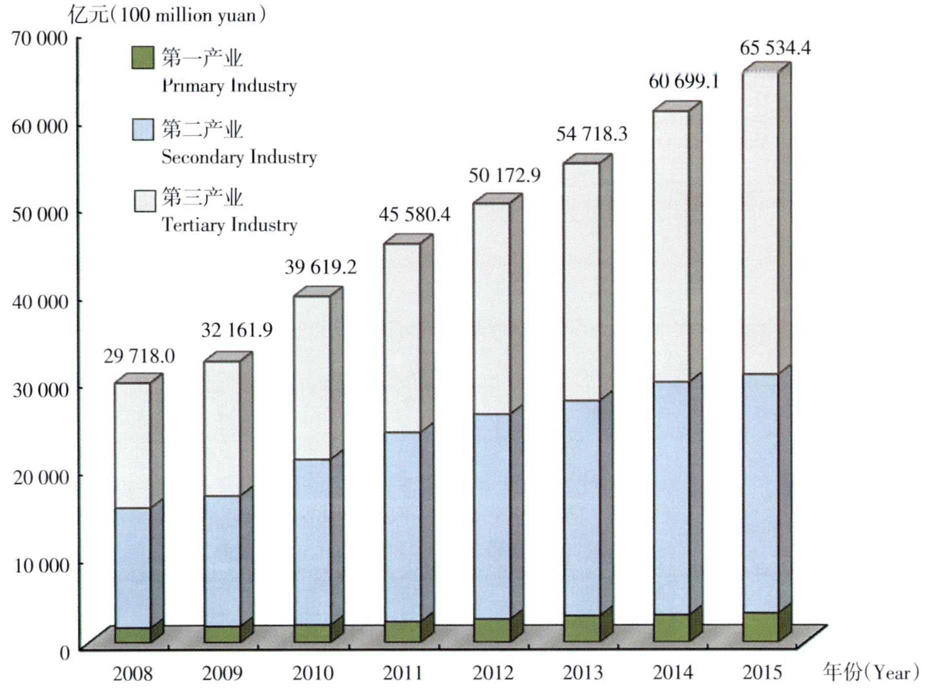

图 2　2015 年全国主要海洋产业增加值构成

Composition of Added Values of the Major Marine Industries in China in 2015

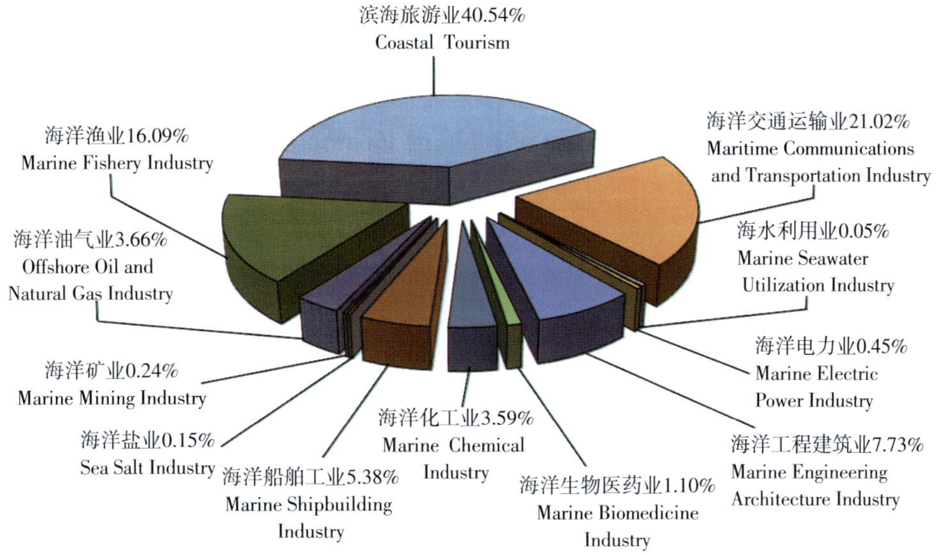

图 3　2015 年沿海地区海洋生产总值

Gross Ocean Product by Coastal Regions in 2015

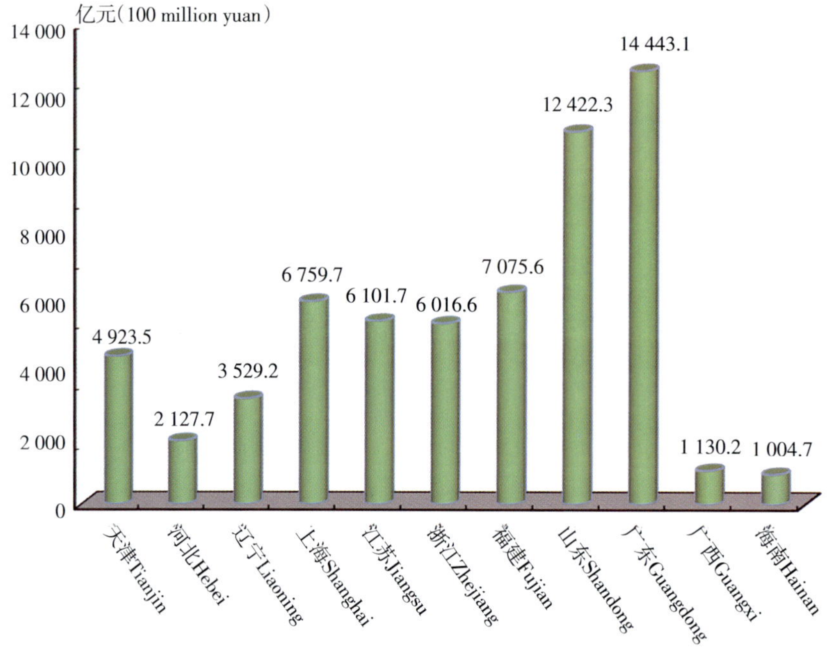

图 4　全国海洋捕捞和海水养殖产量

National Marine Catches and Mariculture Production

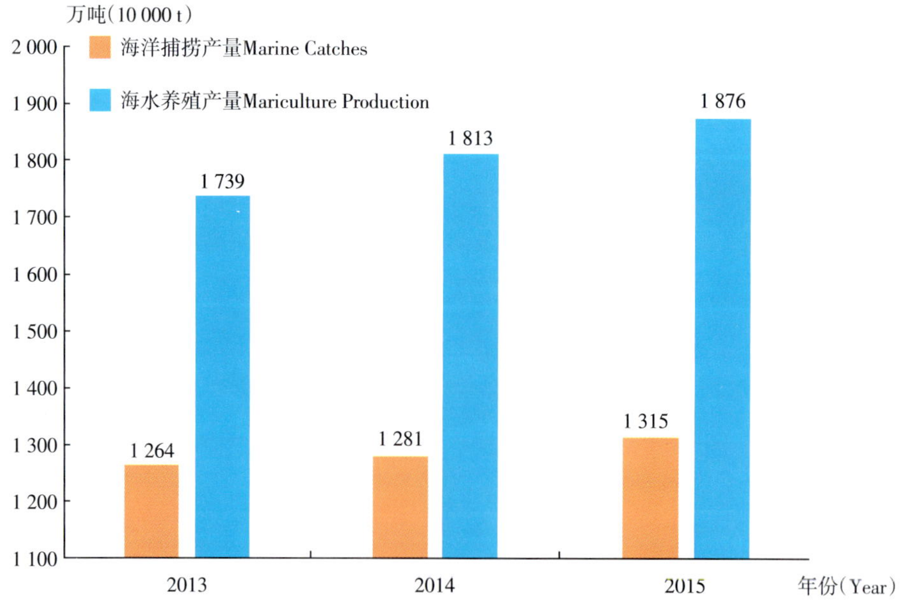

图 5　2015 年沿海地区海洋捕捞和海水养殖产量

Marine Catches and Mariculture Production by Coastal Regions in 2015

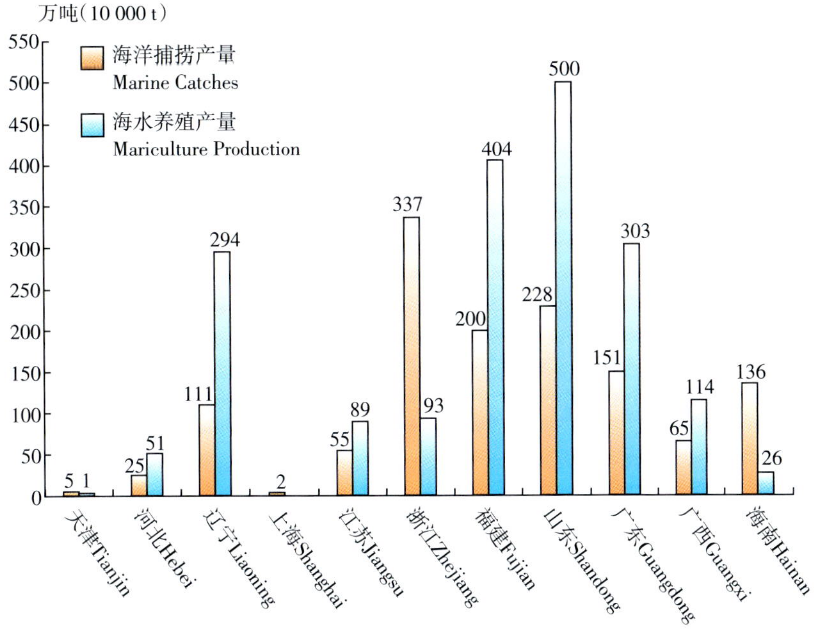

图 6　全国海洋石油和天然气产量

National Output of Offshore Oil and Natural Gas

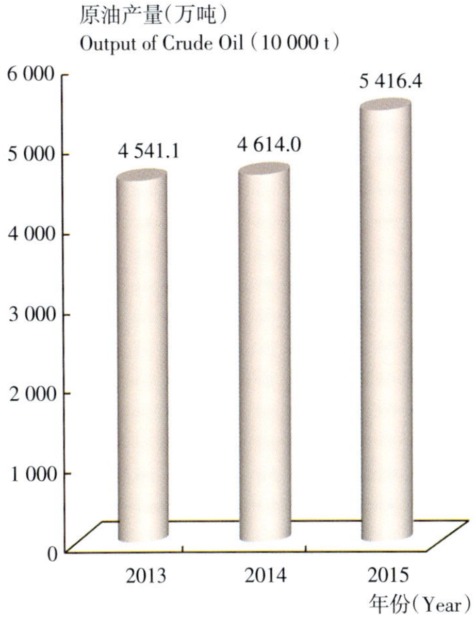

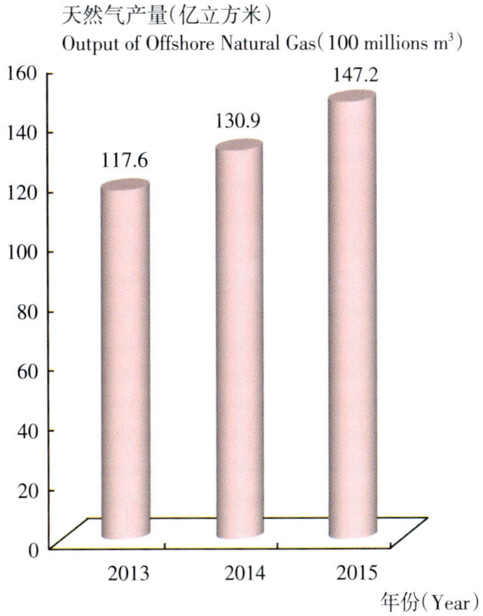

图 7 2015 年海洋矿业产量

Production of Marine Mining Industry in 2015

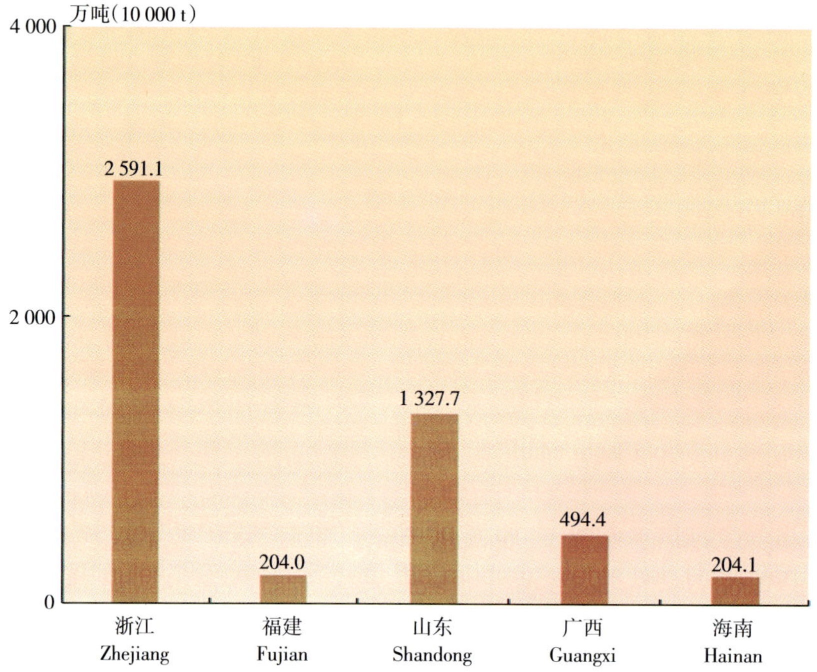

图 8 2015 年沿海地区海盐产量

Sea Salt Production by Coastal Regions in 2015

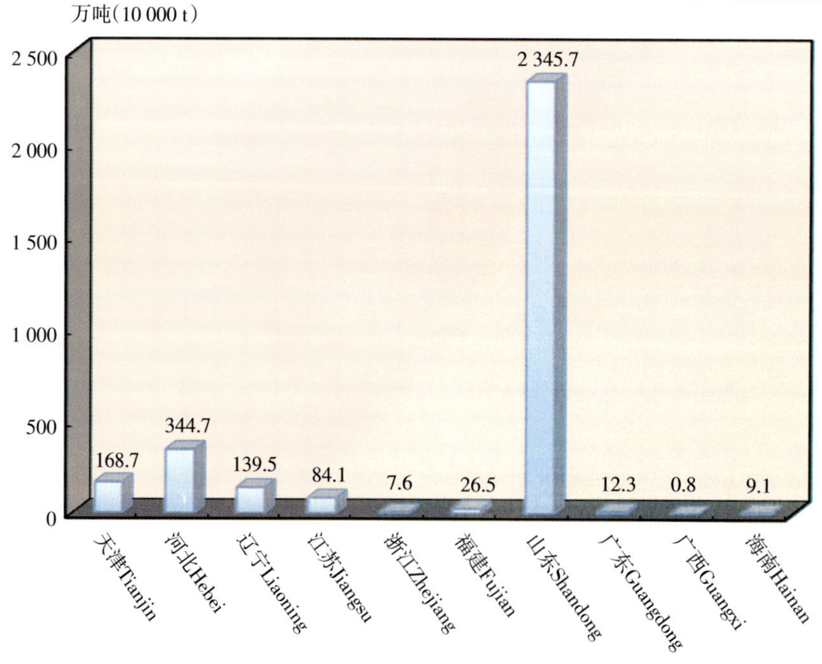

图 9　2015 年沿海地区海洋造船完工量

Completed Quantity of Marine Shipbuilding in the Coastal Region in 2015

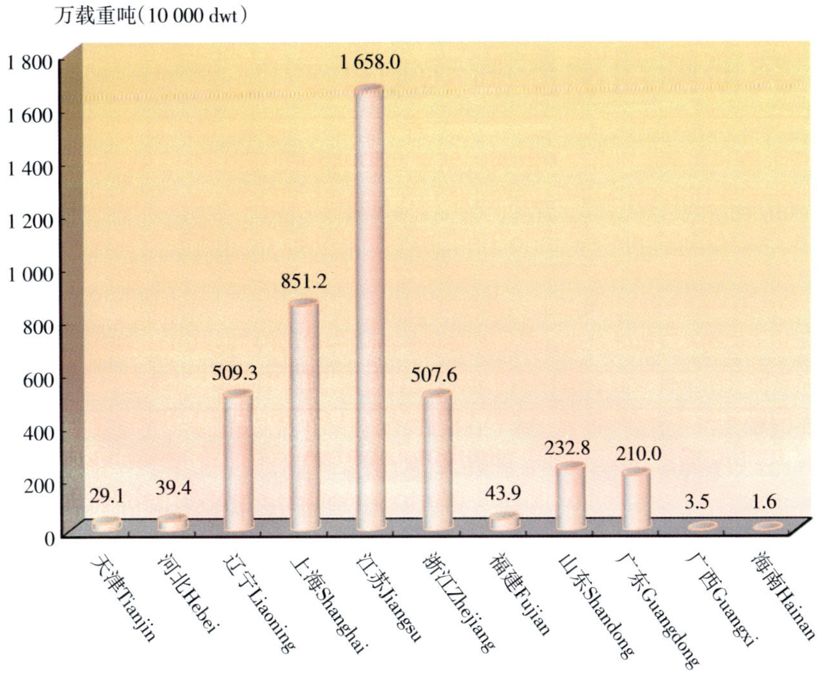

图 10　主要沿海城市接待入境旅游者人数

Number of Oversea Visitor Arrivals in Major Coastal Cities

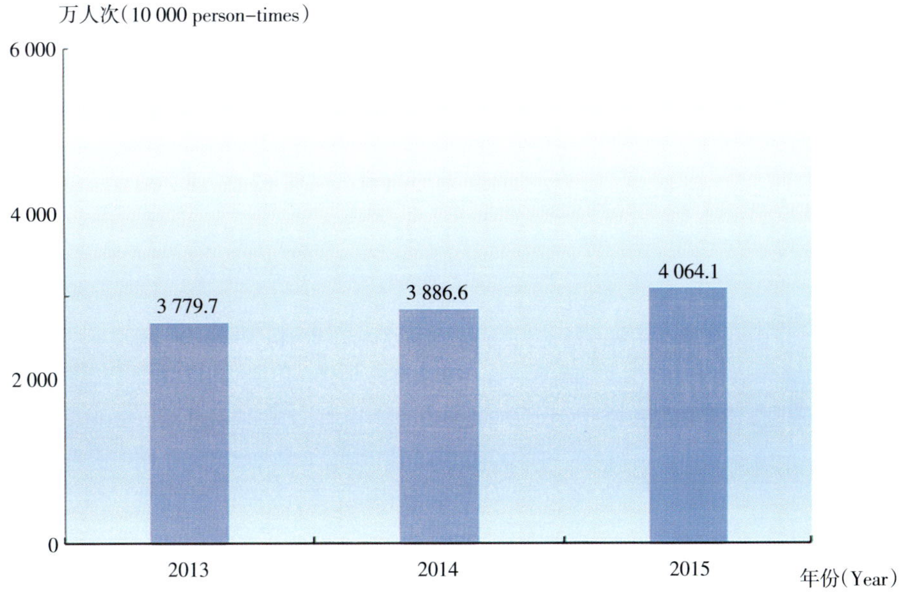

图 11　2015 年沿海地区海洋货物周转量
Maritime Goods Turnover Volume by Coastal Regions in 2015

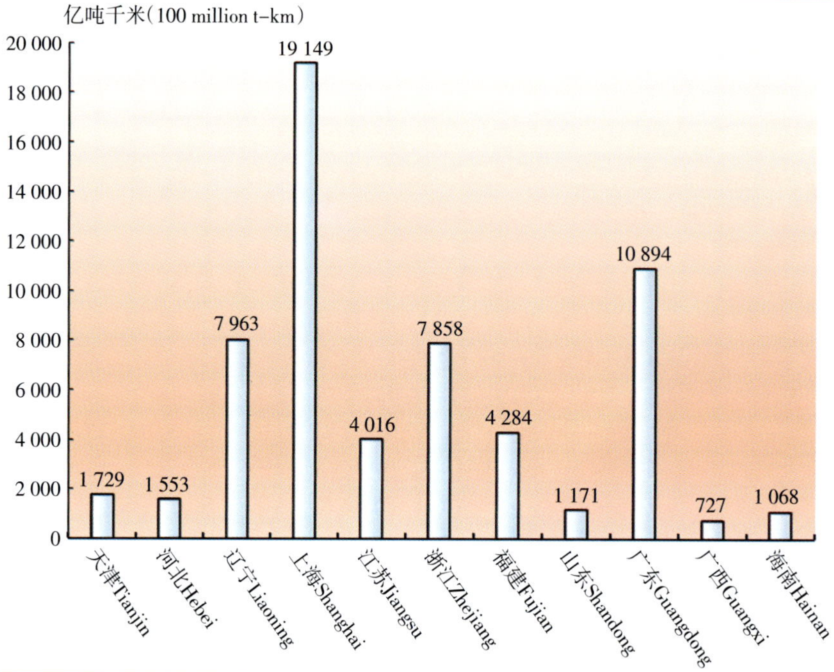

图 12　2015 年沿海港口国际标准集装箱吞吐量
International Standardized Containers Handled by Coastal Seaports in 2015

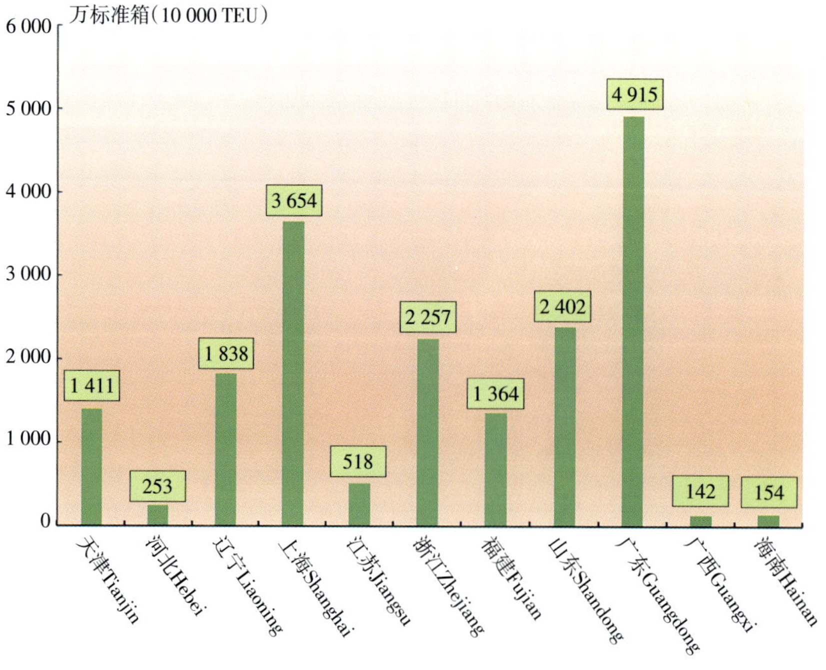

图 13　2015 年三大经济区地区生产总值与海洋生产总值

GDP and GOP in the Three Major Economic Zones in 2015

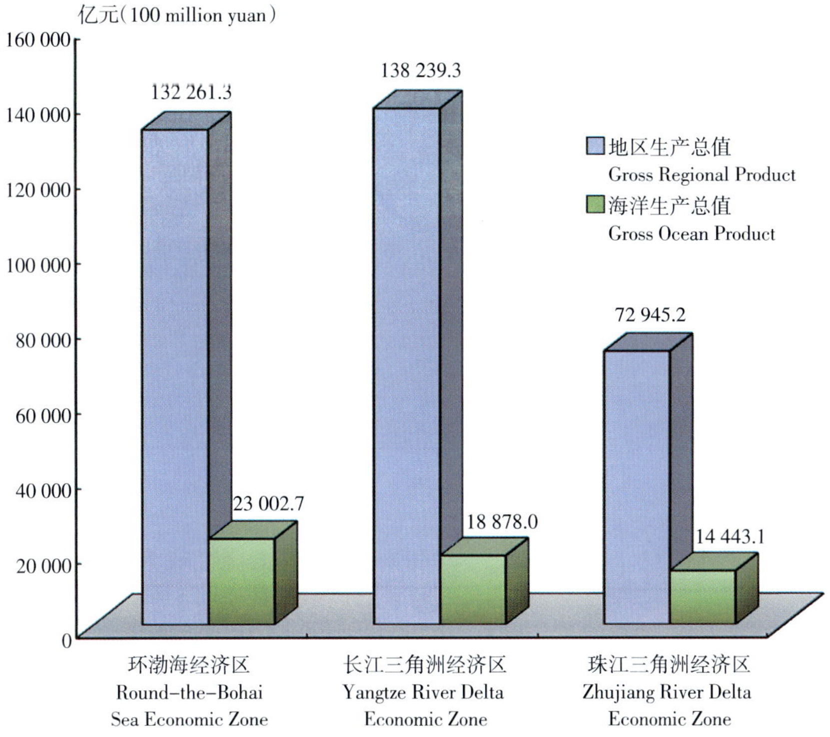

图 14　2015 年三大经济区海洋生产总值占全国海洋生产总值比重

Proportion of the GOP of the Three Major Economic Zones in the National GOP in 2015

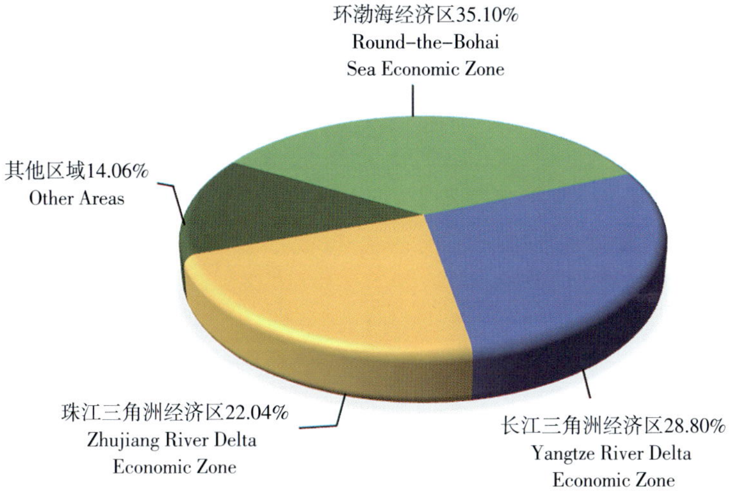

图 15　2015 年沿海地区海洋生产总值
Gross Ocean Product by Coastal Regions in 2015

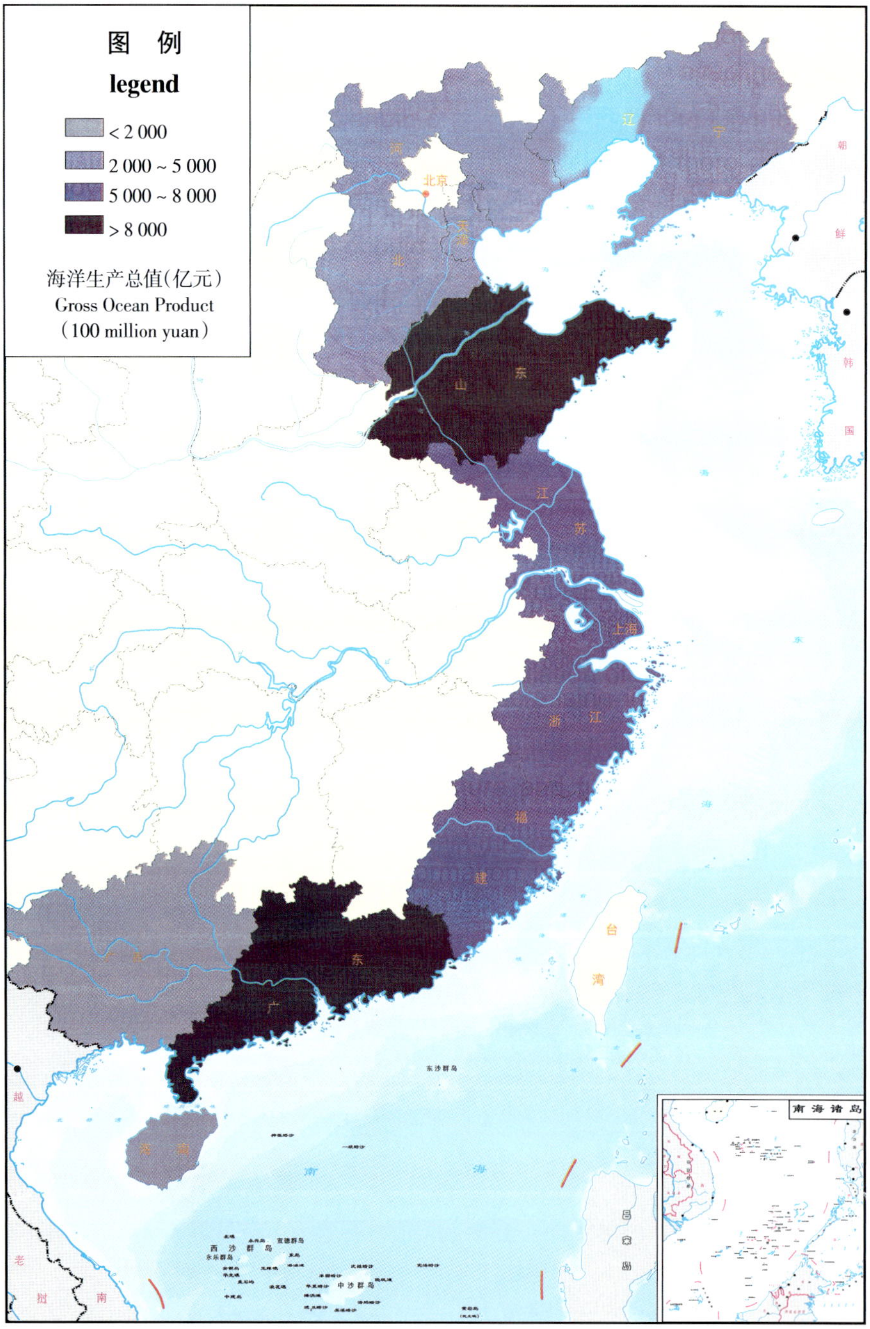

注:香港特别行政区、澳门特别行政区和台湾省未统计在内。
Note: The data of the Hong kong Special Administrative Region, the Macao Special Administrative Region, and Taiwan Province are not included.

图 16　2015 年沿海地区海洋经济贡献

Marine Economic Contributions by Coastal Regions in 2015

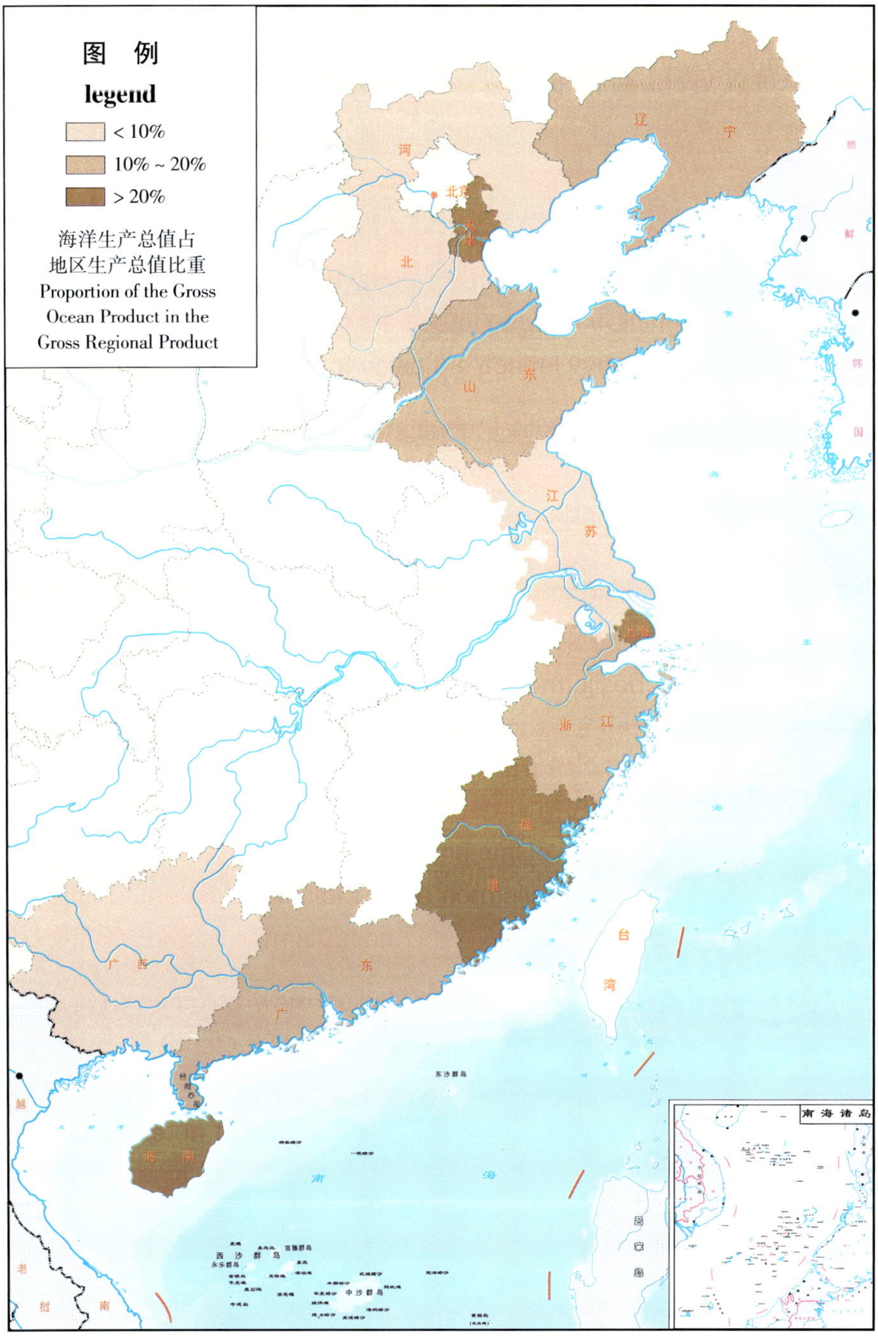

注：香港特别行政区、澳门特别行政区和台湾省未统计在内。

Note: The data of the Hong kong Special Administrative Region, the Macao Special Administrative Region, and Taiwan Province are not included.

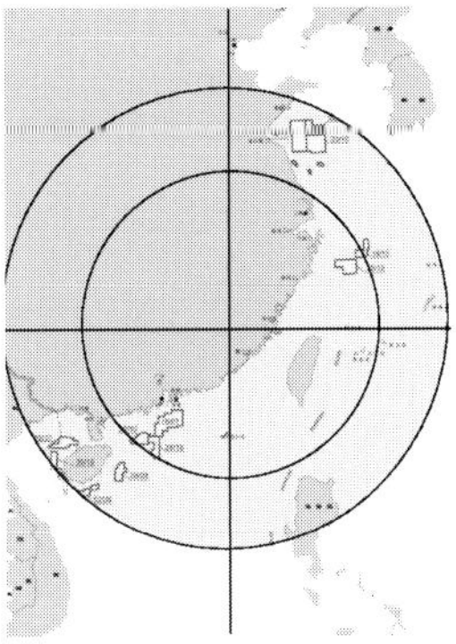

1
综　合　资　料
Integrated Data

1-1 沿海地区行政区划

Administrative Division of Coastal Regions

单位:个 （number）

沿海地区 Coastal Region	沿海城市 Coastal City	沿海地带 Coastal County(District)			
		合　计 Total	县 County	县级市 County-level City	区 District
合　计 Total	**55**	**226**	**56**	**51**	**119**
天　津 Tianjin	1	1			1
河　北 Hebei	3	11	4	1	6
辽　宁 Liaoning	6	22	4	6	12
上　海 Shanghai	1	5	1		4
江　苏 Jiangsu	3	15	7	3	5
浙　江 Zhejiang	7	34	10	9	15
福　建 Fujian	6	33	11	8	14
山　东 Shandong	7	35	5	12	18
广　东 Guangdong	14	46	8	7	31
广　西 Guangxi	3	8	1	1	6
海　南 Hainan	4	16	5	4	7

注：沿海地带中未包括广东省的东莞、中山和海南的三沙、儋州。
Note: The coastal zone does not include the regions under the jurisdiction of Dongguan city, Zhongshan city, Sansha city and Danzhou city.

1-2 沿海行政区划一览表

Table of Administrative Division of Coastal Regions

沿海地区 Coastal Region	地区代码 Zip Code	沿海城市 Coastal City	地区代码 Zip Code	沿海地带 Coastal County (District)	地区代码 Zip Code
天　津 Tianjin	120000			滨海新区 Binhai Xinqu	120116
河　北 Hebei	130000	唐山 Tangshan	130200	丰南区 Fengnan Qu	130207
				曹妃甸区 Caofeidian Qu	130209
				滦南县 Luannan Xian	130224
				乐亭县 Laoting Xian	130225
		秦皇岛 Qinhuangdao	130300	海港区 Haigang Qu	130302
				山海关区 Shanhaiguan Qu	130303
				北戴河区 Beidaihe Qu	130304
				抚宁区 Funing Qu	130306
				昌黎县 Changli Xian	130322
		沧州 Cangzhou	130900	海兴县 Haixing Xian	130924
				黄骅市 Huanghua Shi	130983
辽　宁 Liaoning	210000	大连 Dalian	210200	中山区 Zhongshan Qu	210202
				西岗区 Xigang Qu	210203
				沙河口区 Shahekou Qu	210204
				甘井子区 Ganjingzi Qu	210211
				旅顺口区 Lüshunkou Qu	210212
				金州区 Jinzhou Qu	210213
				普兰店区 Pulandian Qu	210214
				长海县 Changhai Xian	210224
				瓦房店市 Wafangdian Shi	210281
				庄河市 Zhuanghe Shi	210283
		丹东 Dandong	210600	振兴区 Zhenxing Qu	210603
				东港市 Donggang Shi	210681
		锦州 Jinzhou	210700	凌海市 Linghai Shi	210781
		营口 Yingkou	210800	鲅鱼圈区 Bayuquan Qu	210804
				老边区 Laobian Qu	210811
				盖州市 Gaizhou Shi	210881
		盘锦 Panjin	211100	大洼县 Dawa Xian	211121
				盘山县 Panshan Xian	211122
		葫芦岛 Huludao	211400	连山区 Lianshan Qu	211402

1-2 续表1 continued

沿海地区 Coastal Region	地区代码 Zip Code	沿海城市 Coastal City	地区代码 Zip Code	沿海地带 Coastal County (District)	地区代码 Zip Code
				龙港区 Longgang Qu	211403
				绥中县 Suizhong Xian	211421
				兴城市 Xingcheng Shi	211481
上 海 Shanghai	310000			宝山区 Baoshan Qu	310113
				浦东新区 Pudong Xinqu	310115
				金山区 Jinshan Qu	310116
				奉贤区 Fengxian Qu	310120
				崇明县 Chongming Xian	310230
江 苏 Jiangsu	320000	南通 Nantong	320600	通州区 Tongzhou Qu	320612
				海安县 Hai'an Xian	320621
				如东县 Rudong Xian	320623
				启东市 Qidong Shi	320681
				海门市 Haimen Shi	320684
		连云港 Lianyungang	320700	连云区 Lianyun Qu	320703
				赣榆区 Ganyu Qu	320707
				灌云县 Guanyun Xian	320723
				灌南县 Guannan Xian	320724
		盐城 Yancheng	320900	亭湖区 Tinghu Qu	320902
				大丰区 Dafeng Qu	320904
				响水县 Xiangshui Xian	320921
				滨海县 Binhai Xian	320922
				射阳县 Sheyang Xian	320924
				东台市 Dongtai Shi	320981
浙 江 Zhejiang	330000	杭州 Hangzhou	330100	滨江区 Binjiang Qu	330108
				萧山区 Xiaoshan Qu	330109
		宁波 Ningbo	330200	江北区 Jiangbei Qu	330205
				北仑区 Beilun Qu	330206
				镇海区 Zhenhai Qu	330211
				鄞州区 Yinzhou Qu	330212
				象山县 Xiangshan Xian	330225
				宁海县 Ninghai Xian	330226
				余姚市 Yuyao Shi	330281
				慈溪市 Cixi Shi	330282
				奉化市 Fenghua Shi	330283

1-2　续表2 continued

沿海地区 Coastal Region	地区代码 Zip Code	沿海城市 Coastal City	地区代码 Zip Code	沿海地带 Coastal County (District)	地区代码 Zip Code
		温州 Wenzhou	330300	龙湾区 Longwan Qu	330303
				瓯海区 Ouhai Qu	330304
				洞头县 Dongtou Xian	330322
				平阳县 Pingyang Xian	330326
				苍南县 Cangnan Xian	330327
				瑞安市 Rui'an Shi	330381
				乐清市 Yueqing Shi	330382
		嘉兴 Jiaxing	330400	海盐县 Haiyan Xian	330424
				海宁市 Haining Shi	330481
				平湖市 Pinghu Shi	330482
		绍兴 Shaoxing	330600	越城区 Yuecheng Qu	330602
				柯桥区 Keqiao Qu	330603
				上虞区 Shangyu Qu	330604
		舟山 Zhoushan	330900	定海区 Dinghai Qu	330902
				普陀区 Putuo Qu	330903
				岱山县 Daishan Xian	330921
				嵊泗县 Shengsi Xian	330922
		台州 Taizhou	331000	椒江区 Jiaojiang Qu	331002
				路桥区 Luqiao Qu	331004
				玉环县 Yuhuan Xian	331021
				三门县 Sanmen Xian	331022
				温岭市 Wenling Shi	331081
				临海市 Linhai Shi	331082
福　建 Fujian	350000	福州 Fuzhou	350100	马尾区 Mawei Qu	350105
				连江县 Lianjiang Xian	350122
				罗源县 Luoyuan Xian	350123
				平潭县 Pingtan Xian	350128
				福清市 Fuqing Shi	350181
				长乐市 Changle Shi	350182
		厦门 Xiamen	350200	思明区 Siming Qu	350203
				海沧区 Haicang Qu	350205
				湖里区 Huli Qu	350206
				集美区 Jimei Qu	350211

1-2　续表3 continued

沿海地区 Coastal Region	地区代码 Zip Code	沿海城市 Coastal City	地区代码 Zip Code	沿海地带 Coastal County (District)	地区代码 Zip Code
				同安区 Tong'an Qu	350212
				翔安区 Xiang'an Qu	350213
		莆田 Putian	350300	城厢区 Chengxiang Qu	350302
				涵江区 Hanjiang Qu	350303
				荔城区 Licheng Qu	350304
				秀屿区 Xiuyu Qu	350305
				仙游县 Xianyou Xian	350322
		泉州 Quanzhou	350500	丰泽区 Fengze Qu	350503
				泉港区 Quangang Qu	350505
				惠安县 Hui'an Xian	350521
				金门县 Jinmen Xian	350527
				石狮市 Shishi Shi	350581
				晋江市 Jinjiang Shi	350582
				南安市 Nan'an Shi	350583
		漳州 Zhangzhou	350600	云霄县 Yunxiao Xian	350622
				漳浦县 Zhangpu Xian	350623
				诏安县 Zhao'an Xian	350624
				东山县 Dongshan Xian	350626
				龙海市 Longhai Shi	350681
		宁德 Ningde	350900	蕉城区 Jiaocheng Qu	350902
				霞浦县 Xiapu Xian	350921
				福安市 Fu'an Shi	350981
				福鼎市 Fuding Shi	350982
山　东 Shandong	370000	青岛 Qingdao	370200	市南区 Shinan Qu	370202
				市北区 Shibei Qu	370203
				黄岛区 Huangdao Qu	370211
				崂山区 Laoshan Qu	370212
				李沧区 Licang Qu	370213
				城阳区 Chengyang Qu	370214
				胶州市 Jiaozhou Shi	370281
				即墨市 Jimo Shi	370282
		东营 Dongying	370500	东营区 Dongying Qu	370502
				河口区 Hekou Qu	370503

1-2 续表4 continued

沿海地区 Coastal Region	地区代码 Zip Code	沿海城市 Coastal City	地区代码 Zip Code	沿海地带 Coastal County (District)	地区代码 Zip Code
				垦利县 Kenli Xian	370521
				利津县 Lijin Xian	370522
				广饶县 Guangrao Xian	370523
		烟台 Yantai	370600	芝罘区 Zhifu Qu	370602
				福山区 Fushan Qu	370611
				牟平区 Muping Qu	370612
				莱山区 Laishan Qu	370613
				长岛县 Changdao Xian	370634
				龙口市 Longkou Shi	370681
				莱阳市 Laiyang Shi	370682
				莱州市 Laizhou Shi	370683
				蓬莱市 Penglai Shi	370684
				招远市 Zhaoyuan Shi	370685
				海阳市 Haiyang Shi	370687
		潍坊 Weifang	370700	寒亭区 Hanting Qu	370703
				寿光市 Shouguang Shi	370783
				昌邑市 Changyi Shi	370786
		威海 Weihai	371000	环翠区 Huancui Qu	371002
				文登区 Wendeng Qu	371003
				荣成市 Rongcheng Shi	371082
				乳山市 Rushan Shi	371083
		日照 Rizhao	371100	东港区 Donggang Qu	371102
				岚山区 Lanshan Qu	371103
		滨州 Binzhou	371600	沾化区 Zhanhua Qu	371603
				无棣县 Wudi Xian	371623
广 东 Guangdong	440000	广州 Guangzhou	440100	黄埔区 Huangpu Qu	440112
				番禺区 Panyu Qu	440113
				南沙区 Nansha Qu	440115
				增城区 Zengcheng Qu	440118
		深圳 Shenzhen	440300	福田区 Futian Qu	440304
				南山区 Nanshan Qu	440305
				宝安区 Bao'an Qu	440306
				龙岗区 Longgang Qu	440307

1-2 续表5 continued

沿海地区 Coastal Region	地区代码 Zip Code	沿海城市Coastal City	地区代码 Zip Code	沿海地带Coastal County (District)	地区代码 Zip Code
				盐田区 Yantian Qu	440308
		珠海 Zhuhai	440400	香洲区 Xiangzhou Qu	440402
				斗门区 Doumen Qu	440403
				金湾区 Jinwan Qu	440404
		汕头 Shantou	440500	龙湖区 Longhu Qu	440507
				金平区 Jinping Qu	440511
				濠江区 Haojiang Qu	440512
				潮阳区 Chaoyang Qu	440513
				潮南区 Chaonan Qu	440514
				澄海区 Chenghai Qu	440515
				南澳县 Nan'ao Xian	440523
		江门 Jiangmen	440700	蓬江区 Pengjiang Qu	440703
				江海区 Jianghai Qu	440704
				新会区 Xinhui Qu	440705
				台山市 Taishan Shi	440781
				恩平市 Enping Shi	440785
		湛江 Zhanjiang	440800	赤坎区 Chikan Qu	440802
				霞山区 Xiashan Qu	440803
				坡头区 Potou Qu	440804
				麻章区 Mazhang Qu	440811
				遂溪县 Suixi Xian	440823
				徐闻县 Xuwen Xian	440825
				廉江市 Lianjiang Shi	440881
				雷州市 Leizhou Shi	440882
				吴川市 Wuchuan Shi	440883
		茂名 Maoming	440900	电白区 Dianbai Qu	440904
		惠州 Huizhou	441300	惠阳区 Huiyang Qu	441303
				惠东县 Huidong Xian	441323
		汕尾 Shanwei	441500	城　区 Chengqu	441502
				海丰县 Haifeng Xian	441521
				陆丰市 Lufeng Shi	441581
		阳江 Yangjiang	441700	江城区 Jiangcheng Qu	441702
				阳东区 Yangdong Qu	441704
				阳西县 Yangxi Xian	441721

1-2 续表6 continued

沿海地区 Coastal Region	地区代码 Zip Code	沿海城市 Coastal City	地区代码 Zip Code	沿海地带 Coastal County (District)	地区代码 Zip Code
		东莞 Dongguan	441900		
		中山 Zhongshan	442000		
		潮州 Chaozhou	445100	饶平县 Raoping Xian	445122
		揭阳 Jieyang	445200	榕城区 Rongcheng Qu	445202
				揭东区 Jiedong Qu	445203
				惠来县 Huilai Xian	445224
广 西 Guangxi	450000	北海 Beihai	450500	海城区 Haicheng Qu	450502
				银海区 Yinhai Qu	450503
				铁山港区 Tieshangang Qu	450512
				合浦县 Hepu Xian	450521
		防城港 Fangchenggang	450600	港口区 Gangkou Qu	450602
				防城区 Fangcheng Qu	450603
				东兴市 Dongxing Shi	450681
		钦州 Qinzhou	450700	钦南区 Qinnan Qu	450702
海 南 Hainan	460000	海口 Haikou	460100	秀英区 Xiuying Qu	460105
				龙华区 Longhua Qu	460106
				美兰区 Meilan Qu	460108
		三亚 Sanya	460200	海棠区 Haitang Qu	460207
				吉阳区 Jiyang Qu	460203
				天涯区 Tianya Qu	460204
				崖州区 Yazhou Qu	460205
		三沙 Sansha	460300		
		儋州 Danzhou	460400		
		省直辖县 Counties Directly under the Hainan Province Government	469000	琼海市 Qionghai Shi	469002
				文昌市 Wenchang Shi	469005
				万宁市 Wanning Shi	469006
				东方市 Dongfang Shi	469007
				澄迈县 Chengmai Xian	469023
				临高县 Lingao Xian	469024
				昌江黎族自治县 Changjiang Lizu Zizhixian	469026
				乐东黎族自治县 Ledong Lizu Zizhixian	469027
				陵水黎族自治县 Lingshui Lizu Zizhixian	469028

1-3 海洋自然地理
Marine Physical Geography

指　　标	Item	指标值 Data
海洋平均深度（米）	Average Depth of Sea (m)	961
海洋最大深度（米）	Maximum Depth of Sea (m)	5 559
岸线总长度（千米）	Total Length of Coastline (km)	32 000
大陆岸线长度	Length of Continental Coastline	18 000
岛屿岸线长度	Length of Insular Coastline	14 000
>500 m^2岛屿数（个）	Number of Islands >500 m^2 each (unit)	7 300
岛屿面积（万平方千米）	Area of Islands (10 000 km^2)	8

注:海岛数据来源于《全国海岛保护规划》。

Note: Data on islands are derived from the *National Plan for Island Protection*.

1-4 海区海洋石油天然气储量
Offshore Oil and Gas Reserves in the Sea Area

海 区 Sea Area	海洋石油(万吨) Offshore Oil (10 000 t)		海洋天然气(亿立方米) Natural Gas(100 million m^3)	
	累计探明技术可采储量 Proven Technically Recoverable Reserves in the Aggregate	剩余技术可采储量 Surplus Technically Recoverable Reserves	累计探明技术可采储量 Proven Technically Recoverable Reserves in the Aggregate	剩余技术可采储量 Surplus Technically Recoverable Reserves
合 计 Total	**457 990.0**	**60 533.1**	**11 792.2**	**5 103.0**
渤 海 Bohai Sea	332 725.1	46 524.6	2 041.8	514.7
黄 海 Yellow Sea				
东 海 East China Sea	4 126.0	960.7	3 195.5	1 712.5
南 海 South China Sea	121 138.9	13 047.8	6 554.9	2 875.8

注:数据来源于《2015年全国矿产资源储量通报》。

Note: The data come from the *Journal on the National Mineral Resources Reserves in 2015.*

1-5 沿海地区水资源情况
Water Resources by Coastal Regions

地 区 Region	水资源总量（亿立方米） Total Amount of Water Resources (100 million m³)	地表水资源量 Surface Water Resources	地下水资源量 Groundwater Resources	地表水与地下水资源重复量 Duplicate Measurement of Surface Water and Groundwater	人均水资源量（立方米/人） Per Capita Water Resources (m³/person)
全国总计 National Total	**27 962.6**	**26 900.8**	**7 797.0**	**6 735.2**	**2 039.2**
天 津 Tianjin	12.8	8.7	4.9	0.8	83.6
河 北 Hebei	135.1	50.9	113.6	29.4	182.5
辽 宁 Liaoning	179.0	152.0	83.2	56.2	408.1
上 海 Shanghai	64.1	55.3	11.7	2.9	264.8
江 苏 Jiangsu	582.1	462.9	142.4	23.2	730.5
浙 江 Zhejiang	1 407.1	1 390.4	269.8	253.1	2 547.5
福 建 Fujian	1 325.9	1 324.7	332.3	331.1	3 468.7
山 东 Shandong	168.4	84.3	133.1	49.0	171.5
广 东 Guangdong	1 933.4	1 923.4	461.4	451.4	1 792.4
广 西 Guangxi	2 433.6	2 432.2	467.3	465.9	5 096.5
海 南 Hainan	198.2	195.9	50.6	48.3	2 184.9

注：数据来源于《2016中国统计年鉴》。

Note: The data come from the *China Statistical Yearbook 2016*.

1-6 沿海地区湿地面积
Area of Wetlands by Coastal Regions

地　区 Region	湿地总面积 （千公顷） Total Area of Wetlands （1 000 hm^2）	#近海与海岸 Coasts and Seashores
合　计 Total	**12 466.0**	**5 795.9**
天　津 Tianjin	295.6	104.3
河　北 Hebei	941.9	231.9
辽　宁 Liaoning	1 394.8	713.2
上　海 Shanghai	464.6	386.6
江　苏 Jiangsu	2 822.8	1 087.5
浙　江 Zhejiang	1 110.1	692.5
福　建 Fujian	871.0	575.6
山　东 Shandong	1 737.5	728.5
广　东 Guangdong	1 753.4	815.1
广　西 Guangxi	754.3	259.0
海　南 Hainan	320.0	201.7

注：数据来源于《2016中国统计年鉴》。

Note: The data come from the *China Statistical Yearbook 2016*.

1-7　红树林各地类面积
Site Classification and Area of Sharpleaf Mangrove
(*Rhizophora Apiculata*)

单位:公顷　　　　　　　　　　　　　　　　　　　　　　　　　　　　　　　　　(hm²)

地 区 Region	红树林各地类总面积 Total Site Area of Sharpleaf Mangrove	现有面积 Established	未成林面积 Unestablished	宜林地面积 Suitable for Planting
全国总计 National Total	**82 757.2**	**22 024.9**	**1 884.1**	**58 848.2**
浙 江 Zhejiang	5 452.3	20.6	236.1	5 195.6
福 建 Fujian	13 410.1	615.1	286.4	12 508.6
广 东 Guangdong	32 325.9	9 084.0	981.3	22 260.6
广 西 Guangxi	18 029.2	8 374.9	380.3	9 274.0
海 南 Hainan	13 539.7	3 930.3		9 609.4

注：数据来源于国家林业局专项调查。

Note: The data come from Special Survey of State Forestry Administration.

1-8　主要沿海城市气候基本情况
Climate of Major Coastal Cities

城　市 City	年平均气温 （摄氏度） Annual Average Temperature（℃）	年平均相对湿度（%） Annual Average Relative Humidity （%）	全年降水量 （毫米） Annual Precipitation （mm）	全年日照时数 （小时） Annual Sunshine Hours （h）
天　津 Tianjin	13.7	59	574.2	2 221.1
上　海 Shanghai	17.0	74	1 648.8	1 495.6
杭　州 Hangzhou	17.5	75	2 131.9	1 315.8
福　州 Fuzhou	20.7	76	1 778.2	1 319.0
广　州 Guangzhou	22.3	78	2 471.9	1 594.3
海　口 Haikou	25.3	81	1 673.2	2 406.7

注：数据来源于《2016中国统计年鉴》。

Note: The data come from the *China Statistical Yearbook 2016*.

主要统计指标解释

1. 沿海地区　即广义的沿海地区，是指有海岸线(大陆岸线和岛屿岸线)的地区，按行政区划分为沿海省、自治区、直辖市。

2. 沿海城市　是指有海岸线的直辖市和地级市(包括其下属的全部区、县和县级市)。

3. 沿海地带　即狭义的沿海地区，是指有海岸线的县、县级市、区(包括直辖市和地级市的区)。

4. 海洋　是海和洋的统称。洋为地球表面上相连接的广大咸水水体的主体部分。海为地球表面相连接的广大咸水水体被陆地、岛礁、半岛包围或分隔的边缘部分。

5. 水资源总量　指评价区内降水形成的地表和地下产水总量，即地表产流量与降水入渗补给地下水量之和，不包括过境水量。

6. 地表水资源量　指评价区内河流、湖泊、冰川等地表水体中可以逐年更新的动态水量，即当地天然河川径流量。

7. 地下水资源量　指评价区内降水和地表水对饱水岩土层的补给量，包括降水入渗补给量和河道、湖库、渠系、渠灌田间等地表水体的入渗补给量。

8. 地表水与地下水资源重复量　指地表水和地下水相互转化的部分，即天然河川径流量中的地下水排泄量和地下水补给量中来源于地表水的入渗补给量。

9. 湿地　指天然或人工、长久或暂时性的沼泽地、泥炭地或水域地带，包括静止或流动、淡水、半咸水、咸水体，低潮时水深不超过6米的水域以及海岸地带地区的珊瑚滩和海草床、滩涂、红树林、河口、河流、淡水沼泽、沼泽森林、湖泊、盐沼及盐湖。

10. 红树林　指生长在热带、亚热带低能海岸潮间带上部，受周期性潮水浸淹，以红树植物为主体的常绿灌木或乔木组成的潮滩湿地木本生物群落。

11. 气温　指空气的温度，我国一般以摄氏度(℃)为单位表示。气象观测的温度表是放在离地面约1.5米处通风良好的百叶箱里测量的，因此，通常说的气温指的是离地面1.5米处百叶箱中的温度。其统计计算方法为：

月平均气温是将全月各日的平均气温相加，除以该月的天数而得；

年平均气温是将12个月的月平均气温累加后除以12而得。

12. 相对湿度　指空气中实际所含水蒸气密度和同温度下饱和水蒸气密度的百分比值。其统计方法与气温相同。

13. 降水量　指从天空降落到地面的液态或固态(经融化后)水，未经蒸发、渗透、流失而在地面上积聚的深度。其统计计算方法为：

月降水量是将全月各日的降水量累加而得；

年降水量是将12个月的月降水量累加而得。

14. 日照时数　指太阳实际照射地面的时间。其统计方法与降水量相同。

Explanatory Notes on Main Statistical Indicators

1. Coastal Region, i.e., the coastal region in a broad sense, refers to the regions with coastlines (continental and island coastlines), which are divided into the coastal provinces, autonomous regions and municipalities directly under the Central Government according to the administrative zoning.

2. Coastal City refers to the municipalities directly under the Central Government and the prefecture–level cities (including all the districts, counties and county–level cities under them).

3. Coastal Zone, i.e., the coastal region in a narrow sense, refers to the counties, county–level cities and districts with coastlines (including the districts under the municipalities directly under the Central Government and the prefecture–level districts).

4. Ocean is the general name for sea and ocean. Ocean refers to the main body of large salt water connected with the earth surface. Sea refers to the edge areas of the salt water on the earth surface that are compartmentalized or surrounded by land, island, reef or peninsula.

5. Total Water Resources refers to total volume of water resources measured as run–off for surface water from rainfall and recharge for groundwater in a given area, excluding transit water.

6. Surface Water Resources refers to total renewable resources which exist in rivers, lakes, glaciers and other collectors from rainfall and are measured as run–off of rivers.

7. Groundwater Resources refers to replenishment of aquifers with rainfall and surface water.

8. Duplicated Measurement between Surface Water and Groundwater refers to the exchange between surface water and groundwater, i.e. run–off of rivers includes some depletion into groundwater while groundwater includes some replenishment from surface water.

9. Wetlands refer to marshland and peat bog, whether natural or man–made, permanent or temporary; water covered areas, whether stagnant or flowing, with fresh or brackish–fresh or salty water that is less than 6 meters deep at low tide; as well as coral beach, weed beach, mud beach, mangrove, river outlet, rivers, fresh–water marshland, marshland forests, lakes, salty bog and salt lakes along the coastal areas.

10. Mangrove refers to evergreen woody plants or plant communities in tropical or sub–tropical zones which live between the sea and the land in areas which are inundated by tides.

11. Temperature refers to the air temperature. China uses centigrade as the unit. The thermometry used for weather observation is put in a breezy shutter, which is 1.5 meters high from the ground. Therefore, the commonly used temperature refers to the temperature in the breezy shutter 1.5 meters away from the ground. The calculation method is as follows:

Monthly average temperature is the summation of average daily temperature of one month divided by the actual days of that particular month.

Annual average temperature is the summation of monthly averages of a year divided by 12 months.

12. Relative Humidity refers to the ratio of actual water vapour pressure to the saturated water vapour density under the current temperature. The calculation method is the same as that of temperature.

13. Volume of Precipitation refers to the deepness of liquid state or solid state (thawed) water falling from the sky to the ground that has not evaporated, infiltrated or run off. The calculation method is as follows:

Monthly precipitation is the summation of daily precipitation of a month.

Annual precipitation is the summation of 12 months precipitation of a year.

14. Sunshine Hours refer to the actual hours of sun irradiating the earth. The calculation method is the same as that of the precipitation.

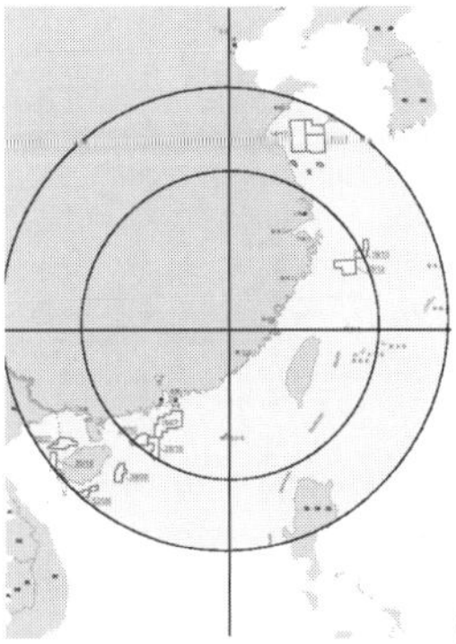

2
海洋经济核算
Marine Economic Accounting

2-1 全国海洋生产总值
National Gross Ocean Product

年 份 Year	海洋生产总值(亿元) Gross Ocean Product (100 million yuan)	第一产业 Primary Industry	第二产业 Secondary Industry	第三产业 Tertiary Industry	海洋生产总值占国内生产总值比重(%) Proportion of the Gross Ocean Product in GDP(%)	海洋生产总值增长速度(%) Growth Rate of the Gross Ocean Product (%)
2001	9 518.4	646.3	4 152.1	4 720.1	8.59	
2002	11 270.5	730.0	4 866.2	5 674.3	9.26	19.8
2003	11 952.3	766.2	5 367.6	5 818.5	8.70	4.2
2004	14 662.0	851.0	6 662.8	7 148.2	9.06	16.9
2005	17 655.6	1 008.9	8 046.9	8 599.8	9.43	16.3
2006	21 592.4	1 228.8	10 217.8	10 145.7	9.84	18.0
2007	25 618.7	1 395.4	12 011.0	12 212.3	9.48	14.8
2008	29 718.0	1 694.3	13 735.3	14 288.4	9.30	9.9
2009	32 161.9	1 857.7	14 926.5	15 377.6	9.21	8.8
2010	39 619.2	2 008.0	18 919.6	18 691.6	9.59	15.3
2011	45 580.4	2 381.9	21 667.6	21 530.8	9.32	10.0
2012	50 172.9	2 670.6	23 450.2	24 052.1	9.28	8.1
2013	54 718.3	3 037.7	24 608.9	27 071.7	9.19	7.8
2014	60 699.1	3 109.5	26 660.0	30 929.6	9.43	7.9
2015	65 534.4	3 327.7	27 671.9	34 534.8	9.51	7.0

注:由于国家实施研发支出核算方法改革,对各年度GDP数据进行了修订,所以海洋生产总值占国内生产总值比重也相应调整。

Note: Because methodology of R&D expenditure accounting is reformed, data of GDP of all years are adjusted, and the data of the proportion of the gross ocean product in GDP are adjusted accordingly.

2-2 全国海洋生产总值构成
Composition of National Gross Ocean Product

单位:% (%)

年 份 Year	第一产业 Primary Industry	第二产业 Secondary Industry	第三产业 Tertiary Industry
2001	6.8	43.6	49.6
2002	6.5	43.2	50.3
2003	6.4	44.9	48.7
2004	5.8	45.4	48.8
2005	5.7	45.6	48.7
2006	5.7	47.3	47.0
2007	5.4	46.9	47.7
2008	5.7	46.2	48.1
2009	5.8	46.4	47.8
2010	5.1	47.8	47.2
2011	5.2	47.5	47.2
2012	5.3	46.7	47.9
2013	5.6	45.0	49.5
2014	5.1	43.9	51.0
2015	5.1	42.2	52.7

2-3 海洋及相关产业增加值
Added Values of Marine and Related Industries

单位:亿元 (100 million yuan)

年 份 Year	合 计 Total	海洋产业 Marine Industry			海洋相关产业 Ocean-related Industries
			主要海洋产业 Major Marine Industry	海洋科研教育管理服务业 Marine Scientific Research, Education, Management and Service	
2001	9 518.4	5 733.6	3 856.6	1 877.0	3 784.8
2002	11 270.5	6 787.3	4 696.8	2 090.5	4 483.2
2003	11 952.3	7 137.7	4 754.4	2 383.3	4 814.6
2004	14 662.0	8 710.1	5 827.7	2 882.5	5 951.9
2005	17 655.6	10 539.0	7 188.0	3 350.9	7 116.6
2006	21 592.4	12 696.7	8 790.4	3 906.4	8 895.6
2007	25 618.7	15 070.6	10 478.3	4 592.3	10 548.0
2008	29 718.0	17 591.2	12 176.0	5 415.2	12 126.8
2009	32 161.9	18 769.4	12 768.4	6 001.0	13 392.5
2010	39 619.2	22 886.4	16 187.8	6 698.5	16 732.8
2011	45 580.4	26 517.6	18 865.2	7 652.4	19 062.8
2012	50 172.9	29 404.6	20 829.9	8 574.8	20 768.2
2013	54 718.3	32 658.7	22 462.3	10 196.4	22 059.6
2014	60 699.1	36 364.9	25 303.4	11 061.5	24 334.1
2015	65 534.4	39 554.9	26 838.8	12 716.0	25 979.5

2-4 海洋及相关产业增加值构成
Composition of the Added Values of Marine and Related Industries

单位:% (%)

年 份 Year	合 计 Total	海洋产业 Marine Industry			海洋相关产业 Ocean-related Industries
			主要海洋产业 Major Marine Industry	海洋科研教育管理服务业 Marine Scientific Research, Education, Management and Service	
2001	100.0	60.2	40.5	19.7	39.8
2002	100.0	60.2	41.7	18.5	39.8
2003	100.0	59.7	39.8	19.9	40.3
2004	100.0	59.4	39.7	19.7	40.6
2005	100.0	59.7	40.7	19.0	40.3
2006	100.0	58.8	40.7	18.1	41.2
2007	100.0	58.8	40.9	17.9	41.2
2008	100.0	59.2	41.0	18.2	40.8
2009	100.0	58.4	39.7	18.7	41.6
2010	100.0	57.8	40.9	16.9	42.2
2011	100.0	58.2	41.4	16.8	41.8
2012	100.0	58.6	41.5	17.1	41.4
2013	100.0	59.7	41.1	18.6	40.3
2014	100.0	59.9	41.7	18.2	40.1
2015	100.0	60.4	41.0	19.4	39.6

2-5 全国主要海洋产业增加值
Added Values of National Major Marine Industries

主要海洋产业 Major Marine Industry	增加值 （亿元） Added Value （100 million yuan）	比上年增长（%） （按可比价计算） Percentage of Increase over Last Year（%） （at comparable price）
合　计 **Total**	**26 838.8**	**8.1**
海洋渔业 Marine Fishery Industry	4 317.4	2.3
海洋油气业 Offshore Oil and Natural Gas Industry	981.9	2.4
海洋矿业 Marine Mining Industry	63.9	7.6
海洋盐业 Sea Salt Industry	41.0	－25.4
海洋船舶工业 Marine Shipbuilding Industry	1 445.7	1.4
海洋化工业 Marine Chemical Industry	964.2	12.4
海洋生物医药业 Marine Biomedicine Industry	295.7	14.0
海洋工程建筑业 Marine Engineering Architecture Industry	2 073.5	13.1
海洋电力业 Marine Electric Power Industry	120.1	13.0
海水利用业 Marine Seawater Utilization Industry	13.7	9.5
海洋交通运输业 Marine Communications and Transportation Industry	5 641.1	7.6
滨海旅游业 Coastal Tourism	10 880.6	11.5

2-6 海洋渔业增加值

Added Value of Marine Fishery Industry

单位:亿元 (100 million yuan)

年 份 Year	增加值 Added Value
2001	966.0
2002	1 091.2
2003	1 145.0
2004	1 271.2
2005	1 507.6
2006	1 672.0
2007	1 906.0
2008	2 228.6
2009	2 440.8
2010	2 851.6
2011	3 202.9
2012	3 560.5
2013	3 997.6
2014	4 126.6
2015	4 317.4

2-7 海洋油气业增加值

Added Value of Offshore Oil and Gas Industry

单位:亿元 (100 million yuan)

年 份 Year	增加值 Added Value
2001	176.8
2002	181.8
2003	257.0
2004	345.1
2005	528.2
2006	668.9
2007	666.9
2008	1 020.5
2009	614.1
2010	1 302.2
2011	1 719.7
2012	1 718.7
2013	1 666.6
2014	1 530.4
2015	981.9

2-8 海洋矿业增加值

Added Value of Marine Mining Industry

单位:亿元 (100 million yuan)

年 份 Year	增加值 Added Value
2001	1.0
2002	1.9
2003	3.1
2004	7.9
2005	8.3
2006	13.4
2007	16.3
2008	35.2
2009	41.6
2010	45.2
2011	53.3
2012	45.1
2013	54.0
2014	59.6
2015	63.9

注:自2008年起部分地区统计矿种增加。

Note: The data from 2008 include added kinds of minerals in some regions.

2-9 海洋盐业增加值

Added Value of Marine Salt Industry

单位:亿元 (100 million yuan)

年 份 Year	增加值 Added Value
2001	32.6
2002	34.2
2003	28.4
2004	39.0
2005	39.1
2006	37.1
2007	39.9
2008	43.6
2009	43.6
2010	65.5
2011	76.8
2012	60.1
2013	63.2
2014	68.3
2015	41.0

2-10 海洋船舶工业增加值
Added Value of Marine Shipbuilding Industry

单位:亿元 (100 million yuan)

年 份 Year	增加值 Added Value
2001	109.3
2002	117.4
2003	152.8
2004	204.1
2005	275.5
2006	339.5
2007	524.9
2008	742.6
2009	986.5
2010	1 215.6
2011	1 352.0
2012	1 291.3
2013	1 213.2
2014	1 395.5
2015	1 445.7

2-11 海洋化工业增加值
Added Value of Marine Chemical Industry

单位:亿元 (100 million yuan)

年 份 Year	增加值 Added Value
2001	64.7
2002	77.1
2003	96.3
2004	151.5
2005	153.3
2006	440.4
2007	506.6
2008	416.8
2009	465.3
2010	613.8
2011	695.9
2012	843.0
2013	813.9
2014	920.0
2015	964.2

注:自2006年起部分地区统计产品品种增加。

Note: The data from 2006 include added kinds of statistical products in some regions.

2-12 海洋生物医药业增加值
Added Value of Marine Biomedicine Industry

单位:亿元 (100 million yuan)

年 份 Year	增加值 Added Value
2001	5.7
2002	13.2
2003	16.5
2004	19.0
2005	28.6
2006	34.8
2007	45.4
2008	56.6
2009	52.1
2010	83.8
2011	150.8
2012	184.7
2013	238.7
2014	258.1
2015	295.7

2-13 海洋工程建筑业增加值
Added Value of Marine Engineering Architecture

单位:亿元 (100 million yuan)

年 份 Year	增加值 Added Value
2001	109.2
2002	145.4
2003	192.6
2004	231.8
2005	257.2
2006	423.7
2007	499.7
2008	347.8
2009	672.3
2010	874.2
2011	1 086.8
2012	1 353.8
2013	1 595.5
2014	1 735.0
2015	2 073.5

2-14 海洋电力业增加值

Added Value of Marine Electric Power Industry

单位:亿元 (100 million yuan)

年 份 Year	增加值 Added Value
2001	1.8
2002	2.2
2003	2.8
2004	3.1
2005	3.5
2006	4.4
2007	5.1
2008	11.3
2009	20.8
2010	38.1
2011	59.2
2012	77.3
2013	91.5
2014	107.7
2015	120.1

2-15 海水利用业增加值

Added Value of Seawater Utilization Industry

单位:亿元 (100 million yuan)

年 份 Year	增加值 Added Value
2001	1.1
2002	1.3
2003	1.7
2004	2.4
2005	3.0
2006	5.2
2007	6.2
2008	7.4
2009	7.8
2010	8.9
2011	10.4
2012	11.1
2013	11.9
2014	12.7
2015	13.7

2-16 海洋交通运输业增加值
Added Value of Marine Communications and Transportation Industry

单位:亿元 (100 million yuan)

年 份 Year	增加值 Added Value
2001	1 316.4
2002	1 507.4
2003	1 752.5
2004	2 030.7
2005	2 373.3
2006	2 531.4
2007	3 035.6
2008	3 499.3
2009	3 146.6
2010	3 785.8
2011	4 217.5
2012	4 752.6
2013	4 876.5
2014	5 336.9
2015	5 641.1

2-17 滨海旅游业增加值
Added Value of Coastal Tourism

单位:亿元 (100 million yuan)

年 份 Year	增加值 Added Value
2001	1 072.0
2002	1 523.7
2003	1 105.8
2004	1 522.0
2005	2 010.6
2006	2 619.6
2007	3 225.8
2008	3 766.4
2009	4 277.1
2010	5 303.1
2011	6 239.9
2012	6 931.8
2013	7 839.7
2014	9 752.8
2015	10 880.6

2-18 沿海地区海洋生产总值
Gross Ocean Product by Coastal Regions

地 区 Region	海洋生产总值(亿元) Gross Ocean Product (100 million yuan)	第一产业 Primary Industry	第二产业 Secondary Industry	第三产业 Tertiary Industry	海洋生产总值占沿海地区生产总值比重(%) Proportion of the Gross Ocean Product in the Gross Regional Product(%)
合 计 Total	**65 534.4**	**3 327.7**	**27 671.9**	**34 534.8**	**16.8**
天 津 Tianjin	4 923.5	15.2	2 803.3	2 105.1	29.8
河 北 Hebei	2 127.7	77.3	986.4	1 064.0	7.1
辽 宁 Liaoning	3 529.2	404.1	1 236.7	1 888.4	15.5
上 海 Shanghai	6 759.7	4.4	2 436.2	4 319.1	26.9
江 苏 Jiangsu	6 101.7	409.5	3 071.5	2 620.7	8.7
浙 江 Zhejiang	6 016.6	462.0	2 164.2	3 390.4	14.0
福 建 Fujian	7 075.6	512.7	2 625.4	3 937.4	27.2
山 东 Shandong	12 422.3	790.0	5 522.4	6 110.0	19.7
广 东 Guangdong	14 443.1	254.0	6 223.3	7 965.9	19.8
广 西 Guangxi	1 130.2	183.1	404.6	542.6	6.7
海 南 Hainan	1 004.7	215.4	198.0	591.3	27.1

2-19 沿海地区海洋生产总值构成
Composition of Gross Ocean Product by Coastal Regions

单位:%　　　　　　　　　　　　　　　　　　　　　　　　　　　　　　　　　　(%)

地　区 Region	海洋生产总值 Gross Ocean Product	第一产业 Primary Industry	第二产业 Secondary Industry	第三产业 Tertiary Industry
合　计 Total	**100.0**	**5.1**	**42.2**	**52.7**
天　津 Tianjin	100.0	0.3	56.9	42.8
河　北 Hebei	100.0	3.6	46.4	50.0
辽　宁 Liaoning	100.0	11.4	35.0	53.5
上　海 Shanghai	100.0	0.1	36.0	63.9
江　苏 Jiangsu	100.0	6.7	50.3	43.0
浙　江 Zhejiang	100.0	7.7	36.0	56.4
福　建 Fujian	100.0	7.2	37.1	55.6
山　东 Shandong	100.0	6.4	44.5	49.2
广　东 Guangdong	100.0	1.8	43.1	55.2
广　西 Guangxi	100.0	16.2	35.8	48.0
海　南 Hainan	100.0	21.4	19.7	58.8

2-20 沿海地区海洋及相关产业增加值

Added Values of Marine and Related Industries by Coastal Regions

单位:亿元 (100 million yuan)

地区 Region	合计 Total	海洋产业 Marine Industry	主要海洋产业 Major Marine Industry	海洋科研教育管理服务业 Industries of Marine Scientific Research, Education, Management and Service	海洋相关产业 Ocean-related Industries
合计 Total	**65 534.4**	**39 554.9**	**26 838.8**	**12 716.0**	**25 979.5**
天津 Tianjin	4 923.5	2 765.0	2 471.2	293.9	2 158.5
河北 Hebei	2 127.7	1 174.2	1 069.5	104.8	953.5
辽宁 Liaoning	3 529.2	2 258.9	1 633.5	625.5	1 270.3
上海 Shanghai	6 759.7	4 076.1	2 220.1	1 856.1	2 661.3
江苏 Jiangsu	6 101.7	3 440.4	2 402.2	1 038.3	2 659.9
浙江 Zhejiang	6 016.6	3 731.5	2 472.5	1 259.0	2 285.1
福建 Fujian	7 075.6	4 082.4	3 118.0	964.4	2 993.2
山东 Shandong	12 422.3	7 515.2	5 239.8	2 275.4	4 907.1
广东 Guangdong	14 443.1	9 085.8	5 140.7	3 945.1	5 357.4
广西 Guangxi	1 130.2	710.6	596.7	113.9	419.6
海南 Hainan	1 004.7	714.7	475.0	239.7	290.0

2-21 沿海地区海洋及相关产业增加值构成
Composition of the Added Values of Marine and Related Industries by Coastal Regions

单位:%　　　　(%)

地 区 Region	合 计 Total	海洋产业 Marine Industry			海洋相关产业 Ocean-related Industries
			主要海洋产业 Major Marine Industry	海洋科研教育管理服务业 Industries of Marine Scientific Research, Education, Management and Service	
合 计 Total	**100.0**	**60.4**	**41.0**	**19.4**	**39.6**
天 津 Tianjin	100.0	56.2	50.2	6.0	43.8
河 北 Hebei	100.0	55.2	50.3	4.9	44.8
辽 宁 Liaoning	100.0	64.0	46.3	17.7	36.0
上 海 Shanghai	100.0	60.3	32.8	27.5	39.7
江 苏 Jiangsu	100.0	56.4	39.4	17.0	43.6
浙 江 Zhejiang	100.0	62.0	41.1	20.9	38.0
福 建 Fujian	100.0	57.7	44.1	13.6	42.3
山 东 Shandong	100.0	60.5	42.2	18.3	39.5
广 东 Guangdong	100.0	62.9	35.6	27.3	37.1
广 西 Guangxi	100.0	62.9	52.8	10.1	37.1
海 南 Hainan	100.0	71.1	47.3	23.9	28.9

主要统计指标解释

1. 海洋经济　是开发、利用和保护海洋的各类产业活动以及与之相关联活动的总和。

2. 海洋生产总值　是海洋经济生产总值的简称，指按市场价格计算的沿海地区常住单位在一定时期内海洋经济活动的最终成果，是海洋产业和海洋相关产业增加值之和。

3. 海洋产业　是开发、利用和保护海洋所进行的生产和服务活动，包括海洋渔业、海洋油气业、海洋矿业、海洋盐业、海洋化工业、海洋生物医药业、海洋电力业、海水利用业、海洋船舶工业、海洋工程建筑业、海洋交通运输业、滨海旅游业等主要海洋产业以及海洋科研教育管理服务业。

4. 海洋科研教育管理服务业　是开发、利用和保护海洋过程中所进行的科研、教育、管理及服务等活动，包括海洋信息服务业、海洋环境监测预报服务、海洋保险与社会保障业、海洋科学研究、海洋技术服务业、海洋地质勘查业、海洋环境保护业、海洋教育、海洋管理、海洋社会团体与国际组织等。

5. 海洋相关产业　是指以各种投入产出为联系纽带，与主要海洋产业构成技术经济联系的上下游产业，涉及海洋农林业、海洋设备制造业、涉海产品及材料制造业、涉海建筑与安装业、海洋批发与零售业、涉海服务业等。

6. 海洋三次产业　我国的海洋三次产业划分如下：

海洋第一产业：是指海洋渔业中的海洋水产品、海洋渔业服务业，以及海洋相关产业中属于第一产业范畴的部门。

海洋第二产业：是指海洋渔业中海洋水产品加工、海洋油气业、海洋矿业、海洋盐业、海洋化工业、海洋生物医药业、海洋电力业、海水利用业、海洋船舶工业、海洋工程建筑业，以及海洋相关产业中属于第二产业范畴的部门。

海洋第三产业：是指除海洋第一、第二产业以外的其他行业。第三产业包括：海洋交通运输业、滨海旅游业、海洋科研教育管理服务业，以及海洋相关产业中属于第三产业范畴的部门。

7. 海洋渔业　包括海水养殖、海洋捕捞、海洋渔业服务业和海洋水产品加工等活动。

8. 海洋油气业　是指在海洋中勘探、开采、输送、加工原油和天然气的生产活动。

9. 海洋矿业　包括海滨砂矿、海滨土砂石与煤矿及深海矿物等的采选活动。

10. 海洋盐业　是指利用海水生产以氯化钠为主要成分的盐产品的活动，包括采盐和盐加工。

11. 海洋船舶工业　是指以金属或非金属为主要材料，制造海洋船舶、海上固定及浮动装置的活动，以及对海洋船舶的修理及拆卸活动。

12. 海洋化工业　包括海盐化工、海水化工、海藻化工及海洋石油化工的化工产品生产活动。

13. 海洋生物医药业　是指以海洋生物为原料或提取有效成分，进行海洋药品与海洋保健品的生产加工及制造活动。

14. 海洋工程建筑业 是指在海上、海底和海岸所进行的用于海洋生产、交通、娱乐、防护等用途的建筑工程施工及其准备活动；包括海港建筑、滨海电站建筑、海岸堤坝建筑、海洋隧道桥梁建筑、海上油气田陆地终端及处理设施建造、海底线路管道和设备安装，不包括各部门、各地区的房屋建筑及房屋装修工程。

15. 海洋电力业 是指在沿海地区利用海洋能、海洋风能进行的电力生产活动。不包括沿海地区的火力发电和核力发电。

16. 海水利用业 是指对海水的直接利用和海水淡化活动，包括利用海水进行淡水生产和将海水应用于工业冷却用水和城市生活用水、消防用水等活动，不包括海水化学资源综合利用活动。

17. 海洋交通运输业 是指以船舶为主要工具从事海洋运输以及为海洋运输提供服务的活动，包括远洋旅客运输、沿海旅客运输、远洋货物运输、沿海货物运输、水上运输辅助活动、管道运输业、装卸搬运及其他运输服务活动。

18. 滨海旅游业 是指以海岸带、海岛及海洋各种自然景观、人文景观为依托的旅游经营、服务活动，主要包括：海洋观光游览、休闲娱乐、度假住宿、体育运动等活动。

Explanatory Notes on Main Statistical Indicators

1. Marine Economy is the summation of various types of industrial activities for developing, utilizing and protecting the ocean as well as the activities associated with there.

2. Gross Ocean Product is the short form of the gross output value of ocean economy, referring to the final result of marine economic activities of the permanent units in the coastal region within a given period calculated at the market price, and the sum total of the added values of the marine industries as the ocean-related industries.

3. Marine Industry refers to the production as service activities for developing, utilizing and protecting the ocean, including major marine industries such as marine fishery industry, offshore oil and gas industry, marine mining industry, marine salt industry, marine chemical industry, marine biomedicine industry, marine electric power industry, seawater utilization industry, marine shipbuilding industry, marine engineering construction industry, marine communications and transportation industry, coastal tourism etc. as well as marine scientific research, education, management and service.

4. Industries of Marine Scientific Research, Education, Management and Service refer to the activities of scientific research, education, management and service carried out in the process of developing, utilizing and protecting the ocean, including marine information service industry, marine environment monitoring and forecasting service, marine insurance and social security industry, marine scientific research, marine technological service industry, ocean geological prospecting industry, marine environmental protection industry, marine education, marine management, marine social organization and international organizations etc.

5. Ocean–Related Industry refers to the lower and upper reaches enterprises that form a technical and economic link with the major marine industries, with various inputs and outputs as ties, involving marine agriculture and forestry, marine equipment manufacturing, ocean–related building and installation industry, marine wholesale and retail industry, ocean–related service industry etc.

6. Marine Three Industries Chinese marine three industries are divided as follows:

Marine primary industry: refers to the marine aquatic products, marine fishery service industry in the marine fishery as well as the sectors belonging to the primary industry category in the ocean–related industries.

Marine secondary industry: refers to the marine aquatic products processing industry in the marine fishery, offshore oil as gas industry, marine mining industry, marine salt industry, marine chemical industry, marine biomedicine industry, marine electric power industry, seawater utilization industry, marine shipbuilding industry, marine engineering construction industry, as well as the sectors belonging to the category of secondary industry in the ocean–related industries.

Marine tertiary Industry: refers to the industries other than the marine primary and secondary industries, including marine communications and transportation industry, coastal tourism, marine scientific research, education, management and service industry as well as the sectors belonging to the category of tertiary industry in the ocean–related industries.

7. Marine Fishery includes mariculture, marine fishing, marine fishery service industry and marine aquatic products processing, etc.

8. Offshore Oil and Gas Industry refers to the production activities of exploring, exploiting, transporting and processing crude oil and natural gas in the ocean.

9. Ocean Mining Industry includes the activities of extracting and dressing beach placers, beach soil and sand, and coal mining and deep–sea mining, etc.

10. Marine Salt Industry refers to the activity of producing the salt products with the sodium chloride as the main component by utilizing seawater, including salt extracting and processing.

11. Shipbuilding Industry refers to the activity of building ocean vessels, offshore fixed and floating equipment with metals or non –metals as main materials as well as repairing and dismantling ocean vessels.

12. Marine Chemical Industry includes the production activities of chemical products of sea salt, seawater, sea algal and marine petroleum chemical industries.

13. Marine Biomedicine Industry refers to the production, processing and manufacturing activities of marine medicines and marine health care products by using marine organisms as raw materials or extracting useful components therefrom.

14. Marine Engineering Building Industry refers to the architectural projects construction and its preparations in the sea, at the sea bottom and seacoast for such uses as marine production, transportation, recreation, protection, etc., including constructions of seaports, coastal power stations, coastal dykes, marine tunnels and bridges, land terminals of offshore oil and gas fields as well as building of processing facilities, and installation of submarine pipelines and equipment, but not the projects of house building and renovation.

15. Marine Electric Power Industry refers to the activities of generating electric power in the coastal region by making use of ocean energies and ocean wind energy. It does not include the thermal and nuclear power generation in the coastal area.

16. Seawater Utilization Industry refers to the activities of the direct use of sea water and the seawater desalination, including those of carrying out the production of desalination and applying the seawater as water for industrial cooling, urban domestic water, water for fire fighting etc., but not the activity of the multipurpose use of seawater chemical resources.

17. Marine Communications and Transportation Industry refers to the activities of carrying out and serving the sea transportations with vessels as main vehicles, including ocean-going passengers transportation, coasta passengers transportation, ocean-going cargo transportation, coastal cargo transportation, auxiliary activities of water transportation, pipeline transportation, loading, unloading and transport as well as other transportation service activities.

18. Coastal Tourism refers to the tourist business and service activities with the backing of coastal zone, sea islands as well as a variety of natural and human landscapes of the ocean, mainly including marine sightseeing, living a life of leisure and recreation, going on vocation and getting accommodation, sports, etc.

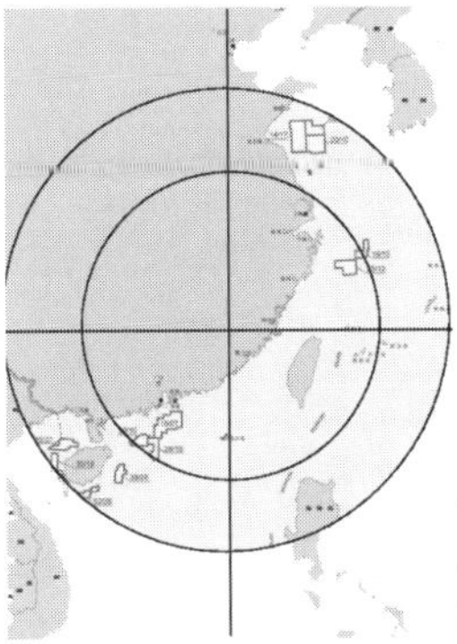

3 主要海洋产业活动

Major Marine Industrial Activities

3-1 全国海洋捕捞和海水养殖产量
National Marine Catches and Mariculture Production

单位:吨 (t)

项 目 Item	2013	2014	2015
远洋捕捞产量 Deep-Sea Fishing Production	**1 351 978**	**2 027 318**	**2 192 000**
海洋捕捞产量 Marine Catches	**12 643 822**	**12 808 371**	**13 147 811**
按品种分 by Species			
鱼类 Fish	8 717 638	8 807 901	9 053 722
甲壳类 Crustacea	2 285 476	2 395 699	2 427 918
贝类 Shellfish	547 556	551 607	555 970
藻类 Algae	28 036	24 299	25 811
头足类 Cephalopoda	664 285	676 715	699 842
其他 Others	400 831	352 150	384 548
海水养殖产量 Mariculture Production	**17 392 453**	**18 126 481**	**18 756 277**
按品种分 by Species			
鱼类 Fish	1 123 576	1 189 667	1 307 628
甲壳类 Crustacea	1 340 218	1 433 763	1 434 917
贝类 Shellfish	12 728 037	13 165 511	13 583 816
藻类 Algae	1 856 804	2 004 576	2 089 153
其他 Others	343 818	332 964	340 763

3-2 沿海地区海洋捕捞和海水养殖产量
Marine Catches and Mariculture Production by Coastal Regions

单位:吨 (t)

地 区 Region	国内海洋捕捞产量 Marine Catches	远洋捕捞产量 Deep-Sea Fishing Production	海水养殖产量 Mariculture Production
全国总计 National Total	**13 147 811**	**2 192 000**	**18 756 277**
天 津 Tianjin	47 094	18 000	10 543
河 北 Hebei	250 447	4 000	506 484
辽 宁 Liaoning	1 107 857	270 000	2 941 965
上 海 Shanghai	16 997	153 000	
江 苏 Jiangsu	554 314	34 000	893 524
浙 江 Zhejiang	3 366 966	570 000	933 431
福 建 Fujian	2 003 917	318 000	4 041 301
山 东 Shandong	2 282 340	469 000	4 995 654
广 东 Guangdong	1 505 126	55 000	3 032 177
广 西 Guangxi	652 028	3 000	1 142 166
海 南 Hainan	1 360 725		259 032

注:远洋捕捞产量全国总计包括北京17 000吨,中农发集团281 000吨。

Note: Data for the national total deep-sea fishing production include 17 000 tons from Beijing and 281 000 tons from the China National Agricultural Development Group Co., Ltd..

3-3 沿海地区海洋原油产量
Output of Offshore Crude Oil by Coastal Regions

单位:万吨 （10 000 t）

地 区 Region		2013	2014	2015
合 计	**Total**	**4 541.09**	**4 613.95**	**5 416.35**
天 津	Tianjin	2 634.68	2 674.09	3 113.82
河 北	Hebei	239.87	237.74	219.84
辽 宁	Liaoning	11.29	48.27	52.99
上 海	Shanghai	18.65	19.53	32.52
山 东	Shandong	291.30	300.30	309.20
广 东	Guangdong	1 345.30	1 334.02	1 687.98

3-4 沿海地区海洋天然气产量
Output of Offshore Natural Gas by Coastal Regions

单位:万立方米 （10 000 m^3）

地 区 Region		2013	2014	2015
合 计	**Total**	**1 176 455**	**1 308 899**	**1 472 400**
天 津	Tianjin	261 682	282 592	289 771
河 北	Hebei	67 201	85 250	77 634
辽 宁	Liaoning	1 410	1 881	1 991
上 海	Shanghai	79 702	88 071	123 977
山 东	Shandong	12 911	12 900	11 676
广 东	Guangdong	753 549	838 205	967 351

3-5 海洋原油出口量及创汇额

Export Volume and Foreign-Exchange Earnings from Offshore Crude Oil by Coastal Regions

单位:万吨,万美元　　　　(10 000 t, 10 000 USD)

地　区 Region	2013		2014		2015	
	出口量 Export Volume	创汇额 Foreign-Exchange Earnings	出口量 Export Volume	创汇额 Foreign-Exchange Earnings	出口量 Export Volume	创汇额 Foreign-Exchange Earnings
合　计 Total	**17**	**11 779**			**9**	**2 946**
天　津 Tianjin	17	11 779			9	2 946

3-6 海洋原油产量、出口量占全国原油产量、出口量比重

Proportion of Offshore Crude Oil Production and Export Volume in the National Total

年　份 Year	海洋原油产量占全国原油产量比重(%) Proportion of Offshore Crude Oil Production in the National Total (%)	海洋原油出口量占全国原油出口量比重(%) Proportion of Offshore Crude Oil Export Volume in the National Total (%)
2001	13.07	46.26
2002	14.40	54.95
2003	15.01	60.77
2004	16.16	83.37
2005	17.51	84.18
2006	17.54	89.53
2007	17.06	71.53
2008	17.96	76.46
2009	19.52	26.61
2010	23.27	24.38
2011	21.94	15.87
2012	21.42	12.43
2013	21.68	10.33
2014	21.82	0.00
2015	25.24	3.07

3-7 沿海地区海洋矿业产量
Output of Marine Mining Industry by Coastal Regions

单位:万吨 (10 000 t)

地 区 Region	产 量 Output		
	2013	2014	2015
合 计 Total	**4 434.2**	**4 727.4**	**4 821.3**
浙 江 Zhejiang	2 134.4	2 243.7	2 591.1
福 建 Fujian	299.5	316.7	204.0
山 东 Shandong	1 318.4	1 507.0	1 327.7
广 西 Guangxi	481.0	454.4	494.4
海 南 Hainan	200.9	205.6	204.1

3-8 沿海地区海盐产量
Output of Sea Salt by Coastal Regions

单位:万吨 （10 000 t）

地 区 Region	海盐产量 Output of Sea Salt		
	2013	2014	2015
合 计 Total	**2 681.1**	**3 085.3**	**3 138.9**
天 津 Tianjin	152.2	161.1	168.7
河 北 Hebei	287.6	336.3	344.7
辽 宁 Liaoning	98.3	128.7	139.5
江 苏 Jiangsu	77.6	81.4	84.1
浙 江 Zhejiang	15.9	8.9	7.6
福 建 Fujian	29.2	38.1	26.5
山 东 Shandong	1 989.9	2 316.6	2 345.7
广 东 Guangdong	7.8	8.4	12.3
广 西 Guangxi	16.1	0.6	0.8
海 南 Hainan	6.6	5.3	9.1

3-9 沿海地区海洋化工产品产量
Output of Marine Chemical Products by Coastal Regions

单位:吨 (t)

地 区 Region	产品产量① Output		
	2013	2014	2015
合 计 Total	**19 095 811**	**22 717 072**	**17 949 463**
天 津 Tianjin	1 641 323	2 940 148	2 910 772
河 北 Hebei	7 942②	17 550②	22 864②
辽 宁 Liaoning	1 021 941	1 162 236	1 151 645
江 苏 Jiangsu	1 933 766	2 523 815	858 614
浙 江 Zhejiang	1 022 884	1 086 436	1 114 938
福 建 Fujian	1 652 964	2 146 419	2 333 389
山 东 Shandong	10 919 468	11 640 468	7 406 541
广 东 Guangdong	895 523	1 200 000	2 150 700

注:①数据为沿海地区部分海洋化工企业产品汇总数据;②为中国盐业总公司数据。

Note: ① The data are collected from the products of part of the chemical enterprises in the coastal region.

② The data are from the China Salt Industry Corporation.

3-10 沿海地区海洋生物医药产品产量

Production of the Marine Biomedicine Industry by Coastal Regions

产品名称 Name	计量单位 Unit	产品产量* Output
海狗油软胶囊 Seal Oil Softgel	万瓶 10 000 bottles	1.84
鲨鱼肝油胶丸 Shark Liver Oil Pill	万粒 10 000 pellets	1 946.78
金枪鱼油胶丸 Tuna Oil Pill	万粒 10 000 pellets	448.74
金枪鱼精制鱼油、降糖多肽等 Refined Tuna Oil, Hypoglycemic Polypeptide, etc.	吨 t	2 280.00
鱼肝油 Cod-Liver Oil	万粒 10 000 pellets	5.22
鱼肝油系列产品 Cod-Liver Oil Series Products	箱 box	2 200.00
鱼肝油乳 Cod-Liver Oil Milk	万瓶 10 000 bottles	102.15
氨基葡萄糖-硫酸软骨素复方片 Glucosamine-Chondroitin Sulfate Compound	吨 t	300.00
甲壳素 Crustaceoxin	吨 t	211.60
角鲨烯胶囊 Houndfish Alkene Capsule	吨 t	235.29
鲨鱼软骨罐头 Shark Cartilage Can	罐 can	700.00
卡拉胶 Carrageenan	吨 t	10 757.46
维生素E胶丸 Vitamin E Capsule	万粒 10 000 pellets	568.56
维生素AD胶丸 Vitamin AD Capsule	吨 t	11 890.00
维生素AD滴剂 Vitamin AD Drops	万瓶 10 000 bottles	144.18
维生素系列产品 Vitamin Series Products	箱 box	1 600.00
琼脂 Agar	吨 t	2 800.00
DHA油 DHA Oil	吨 t	63.00
DHA藻油 DHA Algal Oil	吨 t	66.83
海蛇痹宁胶囊 Sea Snake Anti-rheumatic Capsule	万粒 10 000 pellets	293.65
蚝贝钙片 Oyster Shell Calcium Tablets	万片 10 000 pills	40 880.79
珍珠粉末 Pearl Powder	万袋 10 000 bags	329.32
蚝贝钙片 Oyster Shell Calcium Tablets	万片 10 000 pills	40 880.79

3-10 续表 continued

产品名称 Name	计量单位 Unit	产品产量* Output
八宝惊风散 Babao Infantile Convulsions Powder	万瓶 10 000 bottles	613.84
金牡感冒片 Jinmu Cold Cure Pill	万片 10 000 pills	2 208.56
珠珀惊风散 Zhubo Convulsions Powder	万袋 10 000 bags	404.88
海珠喘息定片 Haizhu Methoxyphenamine Pill	万片 10 000 pills	17 900.12
润科DHA粉剂、油剂 Runke DHA Powder, Oil	吨 t	1 530.00
润科ARA粉剂、油剂 Runke ARA Powder, Oil	吨 t	1 200.00
氨糖胶囊 Glucosamine Capsule	万瓶 10 000 bottles	43.26
壳聚糖胶囊 Chitosan Capsule	万瓶 10 000 bottles	4.31
氨糖软骨素胶囊 Glusosamine Chondroitin Capsule	万瓶 10 000 bottles	3.38
微藻粉 Micro Algae Powder	吨 t	130.67
鲎试剂 Rainbow Limulus Reagent	万支 10 000 bottles	620.00
海蛇酒 Sea Snake Wine	万瓶 10 000 bottles	0.41
珍珠胎囊口服液 Pearl Embryo Oral Liquid	万支 10 000 bottles	500.00
珍珠护肤品 Pearl Cream	万套 10 000 sets	4.00
辅酶Q10 Coenzyme Q10	吨 t	365.00
螺旋藻胶囊 Spirulina Capsule	吨 t	300.00
藻酸双脂钠 Polysacharide Sulfate(PSS)	万片 10 000 pills	282.00
甘糖酯 PGMS	万片 10 000 pills	326.30
海麟舒肝胶囊 Hailin Liver Capsule	万粒 10 000 pellets	27.90
牦牛骨髓蛋白粉 Yak Bone Harrow Protein Powder	万罐 10 000 cans	16.54
伤科接骨片 Bone-Knitting Tablet	吨 t	181.00
藻红蛋白，藻蓝蛋白 Phycoerythrin, Algal Protein	吨 t	5.20

注：数据为沿海地区部分海洋生物医药产品汇总数据。

Note: The data are collected from the products of part of the marine biomedicine enterprices in the coastal region.

3-11 沿海地区海洋修造船完工量

Production of the Marine Shipbuilding Industry by Coastal Regions

部门和地区 Sector and Region	修船完工量(艘) Ships Repaired (unit)	造船完工量 Ships Built	
		艘 Unit	万载重吨 Dead Weight Tonnages (10 000 dwt)
总　计　Total	**15 861**	**1 869**	**5 840.7***
其　中: Including:			
中船工业集团公司 CSSC	691	184	1 172.6
中船重工集团公司 CSIC	654	73	581.7
按地区分: By Regions:			
天　津 Tianjin	226	16	29.1
河　北 Hebei	1 187	15	39.4
辽　宁 Liaoning	210	102	509.3
上　海 Shanghai	1 018	96	851.2
江　苏 Jiangsu	587	317	1 658.0
浙　江 Zhejiang	5 540	373	507.6
福　建 Fujian	2 041	300	43.9
山　东 Shandong	535	34	232.8
广　东 Guangdong	150	26	210.0
广　西 Guangxi	3 000	176	3.5
海　南 Hainan	22	157	1.6

注:*表示造船完工综合吨,单位为万吨。

Note: *The unit is 10 thousand Comprehensive Ton.

3-12 沿海地区海洋货物运输量和周转量

Volume of Maritime Goods Transported and Turnover by Coastal Regions

单位:万吨,亿吨·千米 （10 000 t, 100 million t-km）

地 区 Region	货运量 Volume of Goods Transported	沿 海 Coastal	远 洋 Oceangoing	货物周转量 Volume of Goods Turnover	沿 海 Coastal	远 洋 Oceangoing
全国总计 National Total	**267 667**	**192 982**	**74 685**	**78 460**	**24 224**	**54 236**
天 津 Tianjin	9 850	9 366	484	1 729	1 462	268
河 北 Hebei	4 544	2 868	1 676	1 553	481	1 072
辽 宁 Liaoning	13 439	6 168	7 271	7 963	824	7 139
上 海 Shanghai	47 215	29 070	18 145	19 149	4 000	15 150
江 苏 Jiangsu	22 278	17 397	4 881	4 016	1 920	2 096
浙 江 Zhejiang	54 212	50 336	3 876	7 858	5 672	2 186
福 建 Fujian	25 308	22 863	2 446	4 284	3 477	806
山 东 Shandong	9 542	8 483	1 058	1 171	624	547
广 东 Guangdong	37 841	20 997	16 844	10 894	3 187	7 706
广 西 Guangxi	5 608	5 049	560	727	701	26
海 南 Hainan	10 124	9 543	581	1 068	841	226
其 他 Others	27 706	10 843	16 862	18 049	1 036	17 013

3-13 沿海地区海洋旅客运输量和周转量
Volume of Maritime Passenger Traffic and Turnover by Coastal Regions

单位:万人,亿人·千米 (10 000 persons, 100 million person-km)

地 区 Region	客运量 Passenger Traffic			旅客周转量 Passenger Turnover Volume		
		沿 海 Coastal	远 洋 Oceangoing		沿 海 Coastal	远 洋 Oceangoing
全国总计 National Total	**11 133**	**10 015**	**1 118**	**40.8**	**28.0**	**12.8**
天 津 Tianjin						
河 北 Hebei	5		5	0.3		0.3
辽 宁 Liaoning	504	490	14	6.0	5.3	0.7
上 海 Shanghai	386	385	1	0.8	0.7	0.1
江 苏 Jiangsu	12	5	7	0.6		0.6
浙 江 Zhejiang	2 876	2 876		4.9	4.9	
福 建 Fujian	1 732	1 635	97	2.4	1.9	0.5
山 东 Shandong	1 518	1 421	97	11.4	7.0	4.4
广 东 Guangdong	2 264	1 367	897	9.5	3.4	6.1
广 西 Guangxi	263	263		1.4	1.3	0.1
海 南 Hainan	1 574	1 574		3.4	3.4	

3-14 沿海港口客货吞吐量

Volume of Passenger and Freight Handled at Coastal Seaports

单位:万吨,万人　　　　(10 000 t, 10 000 persons)

地区 Region	货物吞吐量 Cargo Handled	#外贸 Foreign Trade	旅客吞吐量 Passenger Leaving & Arriving	#离港 Leaving
合计 Total	**814 728**	**330 060**	**8 163**	**4 111**
天津 Tianjin	54 051	29 852	52	26
河北 Hebei	91 251	30 752	5	2
辽宁 Liaoning	104 859	23 934	571	288
上海 Shanghai	64 906	37 797	225	113
江苏 Jiangsu	27 332	11 743	7	4
浙江 Zhejiang	109 930	44 183	698	346
福建 Fujian	50 282	20 174	1 012	509
山东 Shandong	134 218	67 856	1 378	671
广东 Guangdong	142 059	48 246	2 867	1 469
广西 Guangxi	20 482	12 556	20	10
海南 Hainan	15 357	2 966	1 328	673

3-15　沿海地区水路国际标准集装箱运量

Volume of International Standardized Containers Traffic by Coastal Regions

单位:万标准箱,万吨　　　　(10 000 TEU, 10 000 t)

地　区 Region	2013		2014		2015	
	箱　数 Number of Containers	重　量 Weight	箱　数 Number of Containers	重　量 Weight	箱　数 Number of Containers	重　量 Weight
合　计 Total	**4 646**	**54 766**	**4 931**	**59 739**	**5 171**	**64 869**
天　津 Tianjin	9	168	28	524	33	626
河　北 Hebei	1	18	6	91	5	71
辽　宁 Liaoning	41	427	42	463	41	411
上　海 Shanghai	1 817	23 722	1 901	25 242	1 675	23 793
江　苏 Jiangsu	508	4 218	475	3 850	464	4 393
浙　江 Zhejiang	239	3 291	272	3 671	292	3 769
福　建 Fujian	345	5 419	440	6 868	512	7 968
山　东 Shandong	120	1 414	117	1 287	74	792
广　东 Guangdong	975	7 244	982	7 462	1 126	9 399
广　西 Guangxi	200	3 080	259	4 410	271	4 767
海　南 Hainan	111	1 874	124	2 949	95	1 567
其　他 Others	280	3 890	285	2 922	583	7 313

3-16 沿海港口国际标准集装箱吞吐量

International Standardized Containers Handled at Coastal Seaports

单位:万标准箱,万吨 (10 000 TEU, 10 000 t)

地 区 Region	2013		2014		2015	
	箱 数 Number of Containers	重 量 Weight	箱 数 Number of Containers	重 量 Weight	箱 数 Number of Containers	重 量 Weight
合 计 Total	**16 967**	**194 694**	**18 179**	**210 056**	**18 907**	**217 788**
天 津 Tianjin	1 301	15 216	1 406	15 904	1 411	15 492
河 北 Hebei	135	2 051	184	2 723	253	3 588
辽 宁 Liaoning	1 798	28 381	1 860	30 834	1 838	31 044
上 海 Shanghai	3 362	34 243	3 529	35 335	3 654	35 850
江 苏 Jiangsu	554	5 536	511	5 074	518	5 091
浙 江 Zhejiang	1 910	19 659	2 136	22 224	2 257	22 914
福 建 Fujian	1 169	14 890	1 271	16 502	1 364	17 754
山 东 Shandong	2 076	22 758	2 256	24 845	2 402	26 851
广 东 Guangdong	4 420	47 947	4 752	51 871	4 915	53 957
广 西 Guangxi	100	1 703	112	1 960	142	2 549
海 南 Hainan	142	2 310	162	2 784	154	2 698

3-17 沿海城市国内旅游人数
Domestic Visitors by Coastal Cities

单位:万人·次 （10 000 person-times）

城市 City		2012	2013	2014
合 计	**Total**	**117 625**	**157 217**	**173 179**
天 津	**Tianjin**			
河 北	**Hebei**	**5 977**	**6 304**	**6 873**
唐 山	Tangshan	2 454	2 770	3 020
秦皇岛	Qinhuangdao	2 313	2 565	2 796
沧 州	Cangzhou	1 210	969	1 057
辽 宁	**Liaoning**	**15 225**	**16 898**	**19 223**
大 连	Dalian	4 687	5 231	5 620
丹 东	Dandong	2 990	2 990	3 469
锦 州	Jinzhou	1 926	3 211	3 545
营 口	Yingkou	1 446	1 620	1 835
盘 锦	Panjin	2 214	1 912	2 299
葫芦岛	Huludao	1 962	1 934	2 455
上 海	**Shanghai**		**25 991**	**26 818**
江 苏	**Jiangsu**	**5 839**	**6 606**	**7 496**
南 通	Nantong	2 408	2 716	3 066
连云港	Lianyungang	1 894	2 136	2 415
盐 城	Yancheng	1 537	1 754	2 015
浙 江	**Zhejiang**	**35 048**	**39 788**	**44 560**
杭 州	Hangzhou	8 237	9 409	10 538
宁 波	Ningbo	5 748	6 226	6 973
温 州	Wenzhou	4 887	5 677	6 358
嘉 兴	Jiaxing	4 101	4 660	5 219
绍 兴	Shaoxing	4 866	5 614	6 287
舟 山	Zhoushan	2 740	3 036	3 400
台 州	Taizhou	4 469	5 166	5 785
福 建	**Fujian**	**11 665**	**12 709**	**14 235**
福 州	Fuzhou	3 107	3 446	3 860
厦 门	Xiamen	2 979	3 411	3 820
莆 田	Putian	1 112	1 590	1 781

3-17 续表 continued

城市 City		2012	2013	2014
泉州	Quanzhou	2 170	1 480	1 658
漳州	Zhangzhou	1 348	1 435	1 608
宁德	Ningde	949	1 347	1 508
山东	**Shandong**	**21 660**	**24 114**	**26 041**
青岛	Qingdao	5 591	6 166	6 659
东营	Dongying	938	1 098	1 185
烟台	Yantai	4 450	4 952	5 348
潍坊	Weifang	4 221	4 702	5 078
威海	Weihai	2 669	2 953	3 189
日照	Rizhao	2 795	3 125	3 375
滨州	Binzhou	996	1 118	1 207
广东	**Guangdong**	**17 411**	**19 338**	**21 695**
广州	Guangzhou	4 017	4 274	4 547
深圳	Shenzhen	2 941	3 352	3 809
珠海	Zhuhai	1 299	1 309	1 516
汕头	Shantou	1 026	1 140	1 276
江门	Jiangmen	1 120	1 240	1 423
湛江	Zhanjiang	1 247	1 392	1 503
茂名	Maoming	399	426	527
惠州	Huizhou	1 122	1 295	1 441
汕尾	Shanwei	518	582	643
阳江	Yangjiang	572	700	876
东莞	Dongguan	1 432	1 481	1 485
中山	Zhongshan	744	808	842
潮州	Chaozhou	438	529	697
揭阳	Jieyang	536	810	1 110
广西	**Guangxi**	**2 811**	**3 260**	**3 807**
北海	Beihai	1 311	1 521	1 771
防城港	Fangchenggang	807	965	1 168
钦州	Qinzhou	693	774	868
海南	**Hainan**	**1 989**	**2 209**	**2 431**
海口	Haikou	935	1 029	1 117
三亚	Sanya	1 054	1 180	1 314

注:数据来源于《2015中国省市经济发展年鉴》。

Note:The data come from *the China Provinces And Cities Economy Development Yearbook* (*2015*).

3-18 主要沿海城市接待入境游客人数
Number of Inbound Tourists Received by Major Coastal Cities

单位:人·次 (person-times)

城　市 City		2013	2014	2015
天　津	Tianjin	758 594	766 326	784 766
秦皇岛	Qinhuangdao	188 041	184 355	157 140
大　连	Dalian	734 605	965 615	984 647
上　海	Shanghai	6 140 911	6 396 150	6 535 887
南　通	Nantong	216 944	187 185	172 999
连云港	Lianyungang	24 228	22 972	20 345
杭　州	Hangzhou	1 060 360	984 701	1 417 404
宁　波	Ningbo	426 258	520 641	723 960
温　州	Wenzhou	290 335	479 898	482 081
福　州	Fuzhou	568 853	569 487	551 572
厦　门	Xiamen	1 072 263	1 170 058	1 273 178
泉　州	Quanzhou	622 476	675 944	663 824
漳　州	Zhangzhou	238 194	270 336	292 426
青　岛	Qingdao	782 632	836 605	880 080
烟　台	Yantai	328 872	358 748	382 398
威　海	Weihai	278 003	300 744	314 224
广　州	Guangzhou	7 681 966	7 832 990	8 035 800
深　圳	Shenzhen	12 148 917	11 825 916	12 187 100
珠　海	Zhuhai	2 632 331	2 913 367	3 095 100
汕　头	Shantou	155 222	176 500	211 500
湛　江	Zhanjiang	190 247	225 500	266 600
中　山	Zhongshan	537 743	602 050	598 300
北　海	Beihai	80 026	74 512	129 053
海　口	Haikou	156 632	136 910	121 975
三　亚	Sanya	481 851	388 637	358 184

3-19 主要沿海城市接待入境游客情况

Breakdown of Inbound Tourists Received by Major Coastal Cities

单位：人·次，人·天 (person-time, night)

城市 City	外国人 Foreigners		香港同胞 Hong Kong Compatriot	
	人次数 Arrivals	人天数 Nights	人次数 Arrivals	人天数 Nights
天　津 Tianjin	690 276	10 896 032	44 576	1 144 671
秦皇岛 Qinhuangdao	145 767	982 435	5 272	30 226
大　连 Dalian	834 868	1 770 788	70 699	142 597
上　海 Shanghai	5 406 936	16 220 808	454 294	1 317 453
南　通 Nantong	151 995	407 396	4 281	9 854
连云港 Lianyungang	16 687	53 356	574	1 195
杭　州 Hangzhou	1 035 644	2 736 084	152 717	391 597
宁　波 Ningbo	569 119	1 127 013	58 875	119 505
温　州 Wenzhou	264 673	797 001	82 685	256 494
福　州 Fuzhou	334 454	1 262 546	65 140	263 924
厦　门 Xiamen	611 115	1 500 037	177 798	405 585
泉　州 Quanzhou	178 750	595 170	341 424	884 237
漳　州 Zhangzhou	81 361	167 240	71 626	146 131
青　岛 Qingdao	633 054	2 628 707	126 976	318 221
烟　台 Yantai	298 733	1 316 104	27 830	103 972
威　海 Weihai	293 324	896 100	2 549	5 937
广　州 Guangzhou	3 079 700	10 388 900	3 895 800	10 485 500
深　圳 Shenzhen	1 646 600	3 531 200	10 062 100	22 170 500
珠　海 Zhuhai	498 100	1 073 400	1 209 200	2 152 700
汕　头 Shantou	113 200	220 700	87 600	149 000
湛　江 Zhanjiang	131 500	327 400	109 700	215 900
中　山 Zhongshan	122 900	416 600	332 600	821 800
北　海 Beihai	65 264	118 030	43 144	76 920
海　口 Haikou	60 350	119 598	18 179	28 466
三　亚 Sanya	233 555	569 415	69 704	145 125

3-19 续表 continued

城市 City	澳门同胞 Macao Compatriot		台湾同胞 Taiwan Compatriot	
	人次数 Arrivals	人天数 Nights	人次数 Arrivals	人天数 Nights
天 津 Tianjin	2 828	75 910	47 086	1 125 862
秦皇岛 Qinhuangdao	230	892	5 871	34 569
大 连 Dalian	2 105	5 063	76 975	144 571
上 海 Shanghai	18 939	58 711	655 718	2 098 298
南 通 Nantong	186	395	16 537	58 098
连云港 Lianyungang	11	22	3 073	12 776
杭 州 Hangzhou	14 141	40 721	214 902	648 875
宁 波 Ningbo	5 476	10 446	90 490	180 338
温 州 Wenzhou	47 499	182 935	87 224	221 171
福 州 Fuzhou	6 645	25 808	145 333	351 964
厦 门 Xiamen	7 280	17 676	476 985	1 112 750
泉 州 Quanzhou	35 933	86 301	107 717	231 914
漳 州 Zhangzhou	6 877	13 225	132 562	205 145
青 岛 Qingdao	28 505	62 712	91 545	257 527
烟 台 Yantai	11 390	35 072	44 445	145 993
威 海 Weihai	623	1 477	17 728	53 932
广 州 Guangzhou	483 200	1 414 600	577 100	1 729 900
深 圳 Shenzhen	55 300	134 000	423 100	970 000
珠 海 Zhuhai	787 800	1 607 500	600 000	1 392 000
汕 头 Shantou	800	1 500	9 900	19 700
湛 江 Zhanjiang	8 100	15 900	17 300	36 700
中 山 Zhongshan	85 400	192 100	57 400	197 200
北 海 Beihai	6 794	12 670	13 851	24 781
海 口 Haikou	1 531	2 030	41 915	58 644
三 亚 Sanya	5 948	11 194	48 977	98 568

主要统计指标解释

1. 海洋捕捞产量 凡是从海洋里捕捞的天然生长的水产品产量为捕捞产量。

2. 海水养殖产量 凡是从人工投放苗种或天然纳苗并进行人工饲养管理的海水养殖水域中捕捞的水产品产量为海水养殖产量。

3. 远洋捕捞产量 由各远洋渔业企业和各生产单位按我国远洋渔业项目管理办法组织的远洋渔船(队)在非我国管辖海域(外国专属经济区水域或公海)捕捞的水产品产量。中外合资、合作渔船捕捞的水产品只统计按协议应属于中方所有的部分。

4. 原油产量 是按净原油量来计算的,能直接用于销售和生产自用的原油量。目前海洋石油系统原油产量计算方法采用倒算法。

原油产量=销售量+期末库存量-期初库存量+海上平台及陆地终端处理厂自用量。

5. 天然气产量 指进入集输管网的销售量和就地利用的全部气量。

天然气产量=外输(销)量+企业自用量

6. 造船综合吨 等于以计量单位载重吨和满载排水量吨的民用船舶的吨位数之和。

7. 货运量 指经船舶实际运送的货物重量。

8. 货物周转量 指实际运送的货物与其运送距离的乘积。

9. 集装箱运量 既包含货重,也包含箱重。箱重系指承运租用的空箱重量凡有运费收入的空箱,其重量应统计为运量,按空箱1吨为货运量1吨计算;若无收入,所承运的空箱一律不作运量统计。

10. 旅客周转量 指实际运送的旅客人数与其运送距离的乘积。

11. 接待人次数 指报告期内我国接待游客人数。游客按出游地分为入境游客和国内游客,按出游时间分为旅游者(过夜游客)和一日游游客(不过夜游客)。

12. 接待人天数 指过夜旅游者的停留天数。

13. 外国人 指外国国籍的人,加入外籍的中国血统华人也计入外国人。

14. 港澳台同胞 指居住在我国香港特别行政区、澳门特别行政区和台湾省的中国同胞。

Explanatory Notes on Main Statistical Indicators

1. Marine Catches refers to the output of the naturally growing aquatic products caught from the sea.

2. Mariculture Production refers to the output of aquatic products whose young are artificially released or naturally collected, and raised and managed artificially, and which are caught from the waters of mariculture.

3. Deep-Sea Fishing Production refers to the output of aquatic products caught in the non-Chinese jurisdictional sea areas (foreign EEE or high sea) by the distant fishing vessels (fleet) organized by various distant fishing businesses and production units according to the management measures of the China distant fishing projects. The aquatic products caught by the Chinese-foreign joint ventures' and cooperative fishing vessels are counted only for the part owned by the Chinese side according to the agreement.

4. Output of Crude Oil is calculated on the basis of the net amount of crude oil, i.e., the amount of crude oil that may be directly used for sale and for the production itself.

Output of crude oil = Volume of sales + Reserves at the end of the period - Reserves at the beginning of the period +Amount for self-use on the platforms and in the terminal processing plants on land.

5. Output of Natural Gas refers to the total gas volume of the sales volume entering the oil collecting and transport pipeline network and that used locally.

Output of natural gas = Volume of sales or transport to other areas + Volume used by the enterprise itself

6. Comprehensive Tonnages of Shipbuilding refers to the sum of tonnage of civilian vessels with the deadweight capacity and full-load displacement as measured.

7. Freight Traffic refers to the weight of cargoes actually transported by vessels.

8. Cargoes Turnover Volume refers to the product of the actually transported cargoes and the transport distance.

9. Freight Volume of Containers includes the weight of both cargoes and container boxes. The weight of container boxes refers to the weight of empty containers rented for transport or having freight income and should be included in the freight volume, one ton of empty boxes equalling to one ton of freight volume. The empty boxes which have no income for transportation are not included in the freight volume.

10. Passenger Turnover Volume refers to the product of the number of passengers actually transported and the shipping distance.

11. Number of Person-Times Received refers to the number of visitors received by China in the period reported. Visitors are divided into inbound visitors and domestic visitors by origin of the travel, and tourists (overnight visitors) and same–day visitors by their length of stay.

12. Number of the Days of Stay refers to the number of the days of stay of tourists.

13. Foreigners refer to the people with foreign nationality, including foreign nationals of Chinese descent.

14. Compatriots from Hong Kong, Macao and Taiwan Province refer to the Chinese compatriots living in the Hong Kong Special Administrative Region, the Macao Special Administrative Region and Taiwan Province.

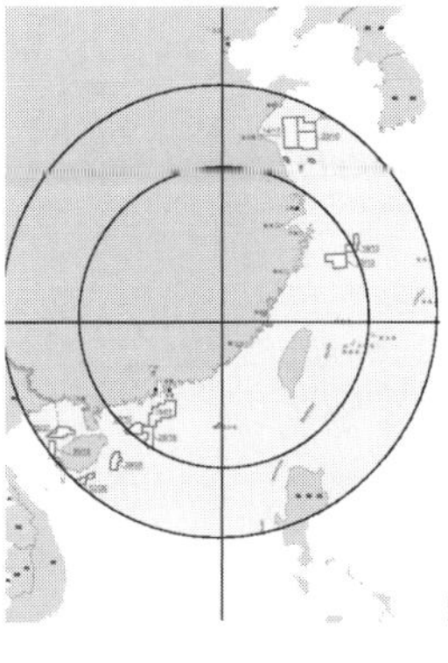

4

主要海洋产业生产能力

Production Capacity of Major Marine Industries

4-1 沿海渔港情况
Coastal Fishing Ports

单位:个 （unit）

地区 Region		合计 Total	中心渔港 Central Fishing Port	一级渔港 Grade1 Fishing Port
合 计	**Total**	**148**	**66**	**82**
天 津	Tianjin			
河 北	Hebei	7	3	4
辽 宁	Liaoning	18	6	12
上 海	Shanghai	1		1
江 苏	Jiangsu	11	6	5
浙 江	Zhejiang	26	10	16
福 建	Fujian	22	9	13
山 东	Shandong	26	14	12
广 东	Guangdong	19	8	11
广 西	Guangxi	8	4	4
海 南	Hainan	10	6	4

资料来源:《2016中国渔业统计年鉴》。省级数据中包含计划单列市数据。

Data Source: *China Fishery Statistical Yearbook 2016*. The data of the provinces include the data of the city specifically designated in the state plan.

4-2 沿海地区海水养殖面积

Mariculture Area by Coastal Regions

单位:公顷 (hm²)

地 区 Region	海水养殖面积 Mariculture Area
合 计 Total	**2 317 763**
天 津 Tianjin	3 165
河 北 Hebei	117 533
辽 宁 Liaoning	933 068
上 海 Shanghai	
江 苏 Jiangsu	181 829
浙 江 Zhejiang	85 881
福 建 Fujian	166 075
山 东 Shandong	563 198
广 东 Guangdong	194 861
广 西 Guangxi	55 015
海 南 Hainan	17 138

4-3 海洋油气勘探情况
Work Volume of Offshore Oil and Gas

地 区 Region	地震测线 Seismic Line		钻井(口) Drilling (well)	
	二维(千米) Two Dimensions (km)	三维(平方千米) Three Dimensions (km^2)	预探井 Wildcat Wells	评价井 Appraisal Wells
合 计 Total	**13 095**	**16 513**	**96**	**79**
天 津 Tianjin		1 780	33	39
其中:合作 Including: Cooporative			3	
河 北 Hebei			19	9
辽 宁 Liaoning				1
上 海 Shanghai	1 017	1 294	3	5
其中:合作 Including: Cooporative			2	
山 东 Shandong	99		1	3
广 东 Guangdong	11 979	13 439	40	22
其中:合作 Including: Cooporative		1 441	4	

4-4 海洋油气生产井情况
Survey of Offshore Oil and Gas Production Wells

单位:口 (well)

地 区 Region	合 计 Total	采油井 Oil Wells	采气井 Gas Wells	注水井 Injection Wells	其他井 Others
合 计 Total	**7 505**	**5 541**	**326**	**1 634**	**4**
天 津 Tianjin	4 023	2 909	127	987	
河 北 Hebei	1 486	1 140	12	334	
辽 宁 Liaoning	360	311	6	43	
上 海 Shanghai	72	20	52		
山 东 Shandong	727	470	8	245	4
广 东 Guangdong	837	691	121	25	

4-5 沿海地区盐田面积和海盐生产能力
Salt Pan Area and Sea Salt Production Capacity by Coastal Regions

地　区 Region	盐田总面积(公顷) Total Area of Salt Pan(hm^2)		生产面积(公顷) Production Area(hm^2)		年末海盐生产能力(万吨) Year-End Capacity of Sea Salt Production(10 000 t)	
	2014	2015	2014	2015	2014	2015
合　计 Total	**402 224**	**349 510**	**295 790**	**263 488**	**3 446.00**	**3 471.92**
天　津 Tianjin	26 923	26 907	26 242	26 221	150.00	160.00
河　北 Hebei	67 992	73 905	58 000	61 587	430.00	419.30
辽　宁 Liaoning	33 664	33 718	28 269	28 319	176.00	180.00
江　苏 Jiangsu	53 200	29 556	11 692	8 341	75.00	70.00
浙　江 Zhejiang	1 971	1 935	1 672	1 631	10.00	8.68
福　建 Fujian	4 116	3 998	4 352	3 711	40.00	27.42
山　东 Shandong	200 890	166 391	157 327	125 646	2 540.00	2 580.00
广　东 Guangdong	8 409	8 580	4 571	4 572	13.00	14.02
广　西 Guangxi	1 600	1 031	900	685	2.00	1.50
海　南 Hainan	3 459	3 489	2 765	2 775	10.00	11.00

4-6 海上风电开工项目
Offshore Wind Power Generation Projects

项 目 Project	开发商 Developer	吊装规模 Lifting and Erecting Size (MW)	总装机规模 Total Installed Capacity (MW)
广东汕头华能海门电厂防波堤风电场 Wave Breaker Wind Farm of Huaneng Haimen Powder Plant, Shantou, Guangdong	华能 China Huaneng Group, Ltd	1.5	1.5
福建省莆田市平海湾50MW近海风电项目 50MW Offshore Wind Power Projects, Putian, Fujian	中闽 Zhongmin Energy Corporation,Ltd	50	50
福建省莆田市南日岛一期400MW近海风电项目 Phase Ⅰ 400MW Offshore Wind Power Projects, Nanri Island, Putian, Fujian	国电龙源 State Grid Longyuan Electric Power Group Corporation, Ltd	12	16
江苏如东C4# C4#, Rudong, Jiangsu	国电龙源 State Grid Longyuan Electric Power Group Corporation, Ltd	100	100
江苏如东10万千瓦潮间带海上风电项目 100 000KW Tidal Zone Wind Farm, Rudong, Jiangsu	中水电 China Hydro Power Corporation,Ltd	80	80
江苏如东海上风电场项目 Offshore Wind Farm, Rudong, Jiangsu	中广核 Zhongguanghe Wind Power Generation Corporation, Ltd	56	152
江苏响水近海风电场项目 Offshore Wind Farm, Xiangshui, Jiangsu	三峡新能源 Three Gorges New Energy Corporation,Ltd	32	148
江苏滨海北区H1# H1#, North Offshore Area, Jiangsu	中电投 China Power Investment Corporation,Ltd	20	100
江苏大丰6MW样机 6MW Model Machine,Dafeng,Jiangsu	天润 Tianrun Energy Corporation,Ltd	6	6
江苏大丰3MW样机 3MW Model Machine,Dafeng,Jiangsu	天润 Tianrun Energy Corporation,Ltd	3	3

4-7 主要潮汐电站分布情况
Distribution of Major Tidal Power Stations

电站名称 Name	运行情况 Status of Operation	装机容量 （千瓦） Installed Capacity （kW）
江厦潮汐试验电站 Jiangxia Experimental Tidal Power Station	1972年开始建造，1980年投入使用，运行至今 It began construction in 1972, was put into use in 1980, and has been in operation so far	4 100
海山潮汐电站 Haishan Tidal Power Station	1972年开始建造，1975年投入使用，运行至今 It began construction in 1972, was put into use in 1975, and has been in operation so far	250
岳浦潮汐电站 Yuepu Tidal Power Station	1970年开始建造，1978年停止运行 It began construction in 1970, and stopped power generation in 1978	150
白沙口潮汐电站 Baishakou Tidal Power Station	1970年开始建造，1978年投入使用，2010年停止运行 It began construction in 1970, was put into use in 1978, and stopped power generation in 2010	960

4-8 主要海上活动船舶

Major Vessels Operating on the Sea

类 别 Type	艘数 （艘） Number of Vessels （unit）	总吨 （万吨位） Gross Tonnage （10 000 t）	净载重量 （万吨） Net Weight Tonnage （10 000 t）	载客量 （客位） Passenger Spaces （seat）	总功率 （千瓦） Total Power （kW）
一、海洋生产用船 Vessels for Marine Production					
海洋渔业船舶 Marine Fishing Vessels	256 821	787.4			15 484 006
远洋渔船 Ocean-going Fishing Vessels	2 512				2 156 612
海洋油气船舶 Offshore Oil and Gas Vessels	199	152.6	90.0	11 122	1 338 793
钻井平台 Drilling Vessels	45	42.8	12.7	5 960	448 378
物探船 Physical Exploration Vessels	12	5.8	1.4	492	57 303
其 他 Others	142	104.0	76.0	4 670	833 112
海洋运输船舶 Marine Transport Vessels	13 410		14 750.3	234 601	39 808 673
二、海洋科研用船 Vessels for Marine Scientific Research					
海洋地质勘探船 Marine Geological Survey Vessels	6	1.2	0.4	219	16 228
海洋调查船 Marine Research Vessels					
中国科学院 Chinese Academy of Sciences	9	2.1	0.2	301	40 544
国家海洋局 State Oceanic Administration	9	4.6		389	37 749

4-9 沿海规模以上港口生产用码头泊位

Berths for Productive Use at the Coastal Seaports above Designed Size

单位：米，个 (m, unit)

港口 Seaport		码头长度 Length of Quay Line	泊位个数 Number of Berths	#万吨级 10 000 Tonnage Class
合　计	**Total**	**735 027**	**5 132**	**1 723**
丹　东	Dandong	7 626	42	25
大　连	Dalian	40 079	222	103
营　口	Yingkou	18 232	83	57
锦　州	Jinzhou	6 119	23	21
秦皇岛	Qinhuangdao	15 928	72	44
黄　骅	Huanghua	8 761	36	30
唐　山	Tangshan	26 121	97	94
天　津	Tianjin	36 473	157	113
烟　台	Yantai	17 911	88	59
威　海	Weihai	4 455	22	14
青　岛	Qingdao	24 608	91	75
日　照	Rizhao	13 683	53	47
上　海	Shanghai	75 161	609	174
连云港	Lianyungang	15 311	62	57
盐　城	Yancheng	8 770	86	15
嘉　兴	Jiaxing	10 933	84	34
宁波-舟山	Ningbo-Zhoushan	82 989	624	157
台　州	Taizhou	12 939	183	9
温　州	Wenzhou	17 093	218	19

4-9 续表 continued

港口 Seaport		码头长度 Length of Quay Line	泊位个数 Number of Berths	#万吨级 10 000 Tonnage Class
福州	Fuzhou	23 687	171	55
莆田	Putian	5 836	45	11
泉州	Quanzhou	15 971	104	25
厦门	Xiamen	28 285	160	71
汕头	Shantou	9 627	87	19
汕尾	Shanwei	1 596	15	2
惠州	Huizhou	9 748	43	21
深圳	Shenzhen	29 021	143	66
虎门	Humen	15 382	117	28
广州	Guangzhou	47 746	514	71
中山	Zhongshan	3 890	55	0
珠海	Zhuhai	17 266	147	27
江门	Jiangmen	10 193	152	3
阳江	Yangjiang	2 232	10	9
茂名	Maoming	2 428	18	9
湛江	Zhanjiang	16 718	143	33
北部湾港	Beibuwan	35 343	256	79
海口	Haikou	6 858	51	15
洋浦	Yangpu	7 955	39	23
八所	Basuo	2 053	10	9

4-10 沿海地区星级饭店基本情况

Star Grade Hotels and Occupancies by Coastal Regions

地 区 Region		饭店数(座) Number of Hotels (unit)	客房数(间) Number of Rooms (unit)	床位数(张) Number of Beds (unit)	客房出租率(%) Room Occupancy (%)
合 计	**Total**	**4 630**	**715 272**	**1 233 147**	
天 津	Tianjin	87	16 059	25 027	51.09
河 北	Hebei	377	54 573	94 078	43.15
辽 宁	Liaoning	382	54 736	118 938	47.31
上 海	Shanghai	228	57 885	87 113	65.80
江 苏	Jiangsu	608	87 187	137 777	58.09
浙 江	Zhejiang	688	104 391	193 765	54.95
福 建	Fujian	352	55 479	87 374	57.86
山 东	Shandong	650	86 371	163 097	53.59
广 东	Guangdong	735	123 109	194 528	54.64
广 西	Guangxi	400	49 954	88 280	54.14
海 南	Hainan	123	25 528	43 170	56.22

4-11 沿海地区旅行社数
Number of Travel Agencies by Coastal Regions

单位:家 (unit)

地 区	Region	旅行社总数 Number of Travel Agencies
合 计	**Total**	**14 184**
天 津	Tianjin	400
河 北	Hebei	1 360
辽 宁	Liaoning	1 253
上 海	Shanghai	1 225
江 苏	Jiangsu	2 160
浙 江	Zhejiang	2 028
福 建	Fujian	846
山 东	Shandong	2 109
广 东	Guangdong	1 901
广 西	Guangxi	539
海 南	Hainan	363

主要统计指标解释

1. **海水养殖面积**　是指利用海上、滩涂、陆基进行鱼、甲壳类(虾、蟹)、贝、藻等海水经济动植物的人工养殖的水面面积。在报告期内无论是否全部收获或尚未收获其产品,均应统计在海水养殖面积中。但有些滩涂、水面不投放苗种或投放少量苗种,只进行一般管理的,不统计为养殖面积。

2. **盐田总面积**　指盐田占有的全部面积。包括储卤、蒸发、保卤、结晶面积、滩内的沟、壕、池、埝、滩坨等面积及滩外的沟、壕、公路及杂地面积。

3. **生产面积**　指直接提供给海盐生产的面积,包括结晶面积、蒸发面积、保卤面积,滩内的沟、壕、池、埝面积及滩坨面积。

4. **年末海盐生产能力**　指年末企业生产原盐的全部设备的综合平衡能力。海盐生产露天作业,受天气影响,因而计算生产能力时,成熟滩田按10年实际平均单位生产面积产量乘以本年成熟滩田生产面积而得,新滩田按设计能力及滩田成熟程度可能达到的产量计算。

5. **海洋渔业船舶**　是指配置机器作为动力的从事海洋渔业生产和辅助渔业生产的船舶。

6. **远洋渔船**　按我国远洋渔业项目管理办法在非我国管辖海域(外国专属经济区水域或公海)进行常年或季节性生产的渔船。

7. **泊位个数**　是指设有系靠船舶设施,在同一时间内可供靠泊最大吨级船舶的艘数。即可靠泊一艘船舶,则计为一个泊位,余类推。泊位分码头泊位和浮筒泊位。

8. **客房数**　指饭店实际可用于接待旅游者的房间数。

9. **床位数**　指饭店实际可用于接待旅游者的床位数。

Explanatory Notes on Main Statistical Indicators

1. Mariculture Area refers to the area of the water surface where seawater economic animals and plants, such as crustacean (shrimp, crab), shellfish and algae, are cultivated at sea, on tidal flat and land. Whether or not all the products in the area have been harvested or the products have not been harvested yet in the period covered by the report, the area is included in the Mariculture Area. But some tidal flats and water surfaces where none or a small amount of the young have been released and only general management is carried out are not included in the Mariculture Area.

2. Total Area of Salt Pans refers to the total area covered by salt pans, including the area for brine storage, evaporation, brine preservation, and crystallization, the area of ditches, moats, ponds and banks within the beach as well as beach mounds, and ditches, moats, highway beyond the beach as well as the area of miscellaneous lands.

3. Area of Salt Pan Production refers to the area directly provided for sea-salt productions, including the area for crystallization, evaporation and brine preservation, the area of ditches, moats, ponds and banks within the beach as well as the area of beach mounds.

4. Year-End Capacity of Crude Salt Production refers to the integrated and balanced capacity of all equipment of the enterprise used for crude salt production at the end of the year. As sea salt production is an open-air operation, which is subject to the effect of weather, the production capacity of a matured salt pan is calculated at the productions of the actual average unit production area in ten years times the production area of the matured salt pan in the current year. The production capacity of new salt pans is calculated at the production that may be reached in the light of the designed capacity and the level of maturity of the salt pan.

5. Marine Fishing Vessels refer to the vessels equipped with machines as motive power and going for marine fishery production and auxiliary fishery production.

6. Deep Sea Fishing Vessels refer to the fishing vessels which carry out production all the year round or seasonally in the non-Chinese jurisdictional, sea areas (foreign EEZ or high sea) according to the China Deep-Sea Fishing Projects Management Measures.

7. Number of Berths refers to the spaces equipped with facilities for docking ships and the number of ships of the maximum tonnage that may dock or anchor in them. A space for a ship to dock is counted as one berth and the rest are reasoned out by analogy. Berths are divided into wharf berths and buoy berths.

8. Number of Rooms refers to the number of guest rooms actually used by the hotels receiving tourists.

9. Number of Beds refers to the number of beds actually used by the hotels receiving tourists.

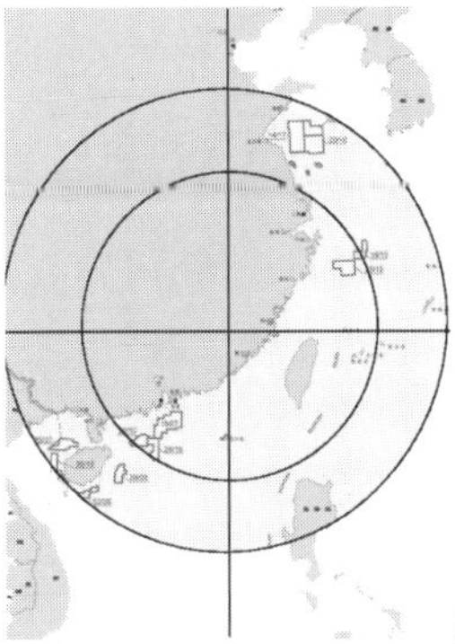

5
涉 海 就 业
Ocean-Related Employment

5-1 全国涉海就业人员情况
Ocean-Related Employed Personnel by Regions

单位：万人 （10 000 persons）

地 区 Region	2001	2014[①]	2015[①]
合 计 Total	**2 107.6**	**3 553.7**	**3 588.5**
天 津 Tianjin	106.4	179.4	181.2
河 北 Hebei	58.0	97.8	98.8
辽 宁 Liaoning	196.0	330.5	333.7
上 海 Shanghai	127.5	215.0	217.1
江 苏 Jiangsu	116.9	197.1	199.0
浙 江 Zhejiang	256.4	432.3	436.6
福 建 Fujian	259.7	437.9	442.2
山 东 Shandong	319.9	539.4	544.7
广 东 Guangdong	505.3	852.0	860.3
广 西 Guangxi	68.9	116.2	117.3
海 南 Hainan	80.6	135.9	137.2
其 他[②] Others	12.0	20.2	20.4

注：① 2014年和2015年为推算数据；② 其他为非沿海地区涉海就业人员数。

Note: ① The data for 2014 and 2015 are the estimated ones;

② Others refer to the number of ocean-related employed persons in the non-coastal regions.

5-2 全国主要海洋产业就业人员情况
Employed Personnel in the Major Marine Industries

单位:万人 （10 000 persons）

海 洋 产 业 Marine Industry	2001	2014	2015
合　计 **Total**	**719.1**	**1 212.5**	**1 224.4**
海洋渔业及相关产业 Marine Fishery and the Related Industries	348.3	587.3	593.0
海洋石油和天然气业 Offshore Oil and Natural Gas Industry	12.4	20.9	21.1
海滨砂矿业 Beach Placer Mining Industry	1.0	1.7	1.7
海洋盐业 Sea Salt Industry	15.0	25.3	25.5
海洋化工业 Marine Chemical Industry	16.1	27.1	27.4
海洋生物医药业 Marine Biomedicine Industry	0.6	1.0	1.0
海洋电力和海水利用业 Marine Electric Power and Seawater Utilization Industry	0.7	1.2	1.2
海洋船舶工业 Marine Shipbuilding Industry	20.6	34.7	35.1
海洋工程建筑业 Marine Engineering Architecture Industry	38.8	65.4	66.1
海洋交通运输业 Maritime Communications and Transportation Industry	50.8	85.7	86.5
滨海旅游业 Coastal Tourism	78.3	132.0	133.3
其他海洋产业 Other Marine Industries	136.5	230.2	232.4

注:2014年和2015年为推算数据。

Note: The data for 2014 and 2015 are the estimated ones.

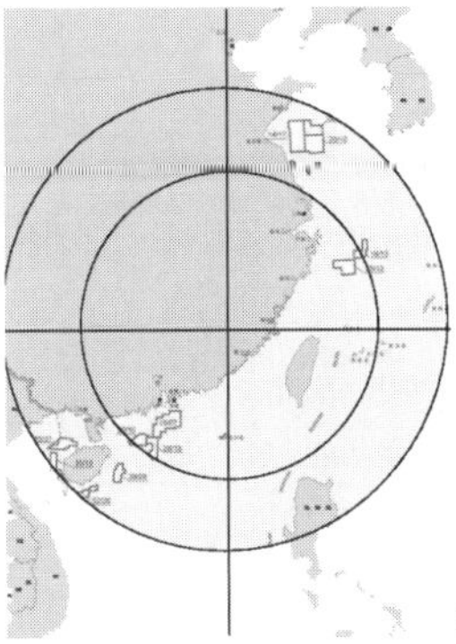

6 海洋科学技术 Marine Science and Technology

6-1 分行业海洋科研机构及人员情况
Marine Scientific Research Institutions and Personnel by Industry

行 业 Industry	机构数(个) Number of Institutions(unit)	从业人员(人) Employees(person)
合 计 **Total**	**192**	**42 331**
海洋基础科学研究 **Marine Basic Scientific Research**	**103**	**18 802**
海洋自然科学 Marine Natural Science	61	14 500
海洋社会科学 Marine Social Science	3	784
海洋农业科学 Marine Agricultural Science	37	3 427
海洋生物医药 Marine Biomedicine	2	91
海洋工程技术研究 **Marine Engineering Technology Research**	**74**	**19 364**
海洋化学工程技术 Marine Chemical Engineering Technology	12	6 192
海洋生物工程技术 Marine Bioengineering Technology	2	222
海洋交通运输工程技术 Marine Communications and Transport Technology	19	5 717

6-1 续表 continued

行 业 Industry	机构数(个) Number of Institutions (unit)	从业人员(人) Employees (person)
海洋能源开发技术 Marine Energies Development Technology	4	1 655
海洋环境工程技术 Marine Environmental Engineering Technology	14	1 183
河口水利工程技术 Estuarine Water Conservancy Engineering Technology	19	3 490
其他海洋工程技术 Other Marine Engineering Technology	4	905
海洋信息服务业 **Marine Information Service**	**9**	**990**
其他海洋信息服务 Other Marine Information Services	9	990
海洋技术服务业 **Marine Technological Service Industry**	**6**	**3 175**
海洋工程管理服务 Marine Engineering Management Service	1	546
其他海洋专业技术服务 Other Marine Professional and Technological Services	5	2 629

注:机构为县级以上科研机构;行业分类参照《海洋及相关产业分类》标准。

Note: The institutions are the scientific research institutions above the county level; The classification of industries follows the standard *Classification of Marine Industries and the Related Industries*.

6-2　分行业海洋科研机构科技活动人员学历构成

Educational Background Composition of the Personnel Engaged in Scientific and Technological Activities in Marine Scientific Research Institutions by Industry

单位:人　　　　(person)

行　业 Industry	科技活动人员 Personnel Engaged in Scientific Activities	博士 Doctor	硕士 Master	大学生 Graduate	大专生 College Graduate
合　计 **Total**	**35 860**	**8 749**	**11 793**	**10 417**	**2 905**
海洋基础科学研究 **Marine Basic Scientific Research**	**17 244**	**5 921**	**5 016**	**4 128**	**1 179**
海洋自然科学 Marine Natural Science	13 523	5 263	3 834	2 808	826
海洋社会科学 Marine Social Science	750	136	299	255	59
海洋农业科学 Marine Agricultural Science	2 899	521	874	1 017	283
海洋生物医药 Marine Biomedicine	72	1	9	48	11
海洋工程技术研究 **Marine Engineering Technology Research**	**14 716**	**2 184**	**4 902**	**5 231**	**1 546**
海洋化学工程技术 Marine Chemical Engineering Technology	5 166	891	1 395	1 614	681
海洋生物工程技术 Marine Bioengineering Technology	199	18	107	51	23

6-2 续表 continued

行 业 Industry	科技活动人员 Personnel Engaged in Scientific Activities	博士 Doctor	硕士 Master	大学生 Graduate	大专生 College Graduate
海洋交通运输工程技术 Marine Communications and Transport Technology	3 461	212	1 372	1 396	332
海洋能源开发技术 Marine Energies Development Technology	1 546	368	473	461	230
海洋环境工程技术 Marine Environmental Engineering Technology	1 050	63	306	574	90
河口水利工程技术 Estuarine Water Conservancy Engineering Technology	2 622	530	883	959	169
其他海洋工程技术 Other Marine Engineering Technology	672	102	366	176	21
海洋信息服务业 **Marine Information Service**	**951**	**93**	**390**	**308**	**88**
其他海洋信息服务 Other Marine Information Services	951	93	390	308	88
海洋技术服务业 **Marine Technological Service Industry**	**2 949**	**551**	**1 485**	**750**	**92**
海洋工程管理服务 Marine Engineering Management Service	535	39	275	206	15
其他海洋专业技术服务 Other Marine Professional and Technological Services	2 414	512	1 210	544	77

6-3 分行业海洋科研机构科技活动人员职称构成
Professional Title Composition of Personnel Engaged in Scientific and Technological Activities in the Marine Scientific Research Institutions by Industry

单位：人 (person)

行 业 Industry	科技活动人员 Personnel Engaged in Scientific Activities	高级职称 Senior Professional Title	中级职称 Mid-level Professional Title	初级职称 Junior Professional Title
合 计 Total	**35 860**	**14 703**	**11 970**	**5 460**
海洋基础科学研究 Marine Basic Scientific Research	**17 244**	**7 379**	**6 163**	**2 236**
海洋自然科学 Marine Natural Science	13 523	5 972	4 936	1 617
海洋社会科学 Marine Social Science	750	367	197	62
海洋农业科学 Marine Agricultural Science	2 899	1 034	1 011	522
海洋生物医药 Marine Biomedicine	72	6	19	35
海洋工程技术研究 Marine Engineering Technology Research	**14 716**	**6 184**	**4 634**	**2 102**
海洋化学工程技术 Marine Chemical Engineering Technology	5 166	2 242	1 544	606
海洋生物工程技术 Marine Bioengineering Technology	199	60	81	57

6-3 续表 continued

行 业 Industry	科技活动人员 Personnel Engaged in Scientific Activities	高级职称 Senior Professional Title	中级职称 Mid-level Professional Title	初级职称 Junior Professional Title
海洋交通运输工程技术 Marine Communications and Transport Technology	3 461	1 059	1 166	791
海洋能源开发技术 Marine Energies Development Technology	1 546	742	498	93
海洋环境工程技术 Marine Environmental Engineering Technology	1 050	411	373	210
河口水利工程技术 Estuarine Water Conservancy Engineering Technology	2 622	1 341	718	282
其他海洋工程技术 Other Marine Engineering Technology	672	329	254	63
海洋信息服务业 Marine Information Service	**951**	**336**	**298**	**210**
其他海洋信息服务 Other Marine Information Services	951	336	298	210
海洋技术服务业 Marine Technological Service Industry	**2 949**	**804**	**875**	**912**
海洋工程管理服务 Marine Engineering Management Service	535	218	243	56
其他海洋专业技术服务 Other Marine Professional and Technological Services	2 414	586	632	856

6-4　分行业海洋科研机构经费收入

Routine Fund Receipts of the Marine Scientific Research Institutions by Industry

单位:千元　　　　(1 000 yuan)

行　业 Industry	经费收入总额 Fund Total	本年收入合计 Total Annual Income	基本建设中政府投资 Government Investment in the Capital Construction
合　计 Total	**33 334 007**	**31 020 439**	**2 312 818**
海洋基础科学研究 Marine Basic Scientific Research	**16 796 756**	**14 929 214**	**1 867 542**
海洋自然科学 Marine Natural Science	14 323 935	12 677 194	1 646 741
海洋社会科学 Marine Social Science	470 242	435 186	35 056
海洋农业科学 Marine Agricultural Science	1 983 515	1 797 770	185 745
海洋生物医药 Marine Biomedicine	19 064	19 064	0
海洋工程技术研究 Marine Engineering Technology Research	**13 925 429**	**13 503 275**	**421 404**
海洋化学工程技术 Marine Chemical Engineering Technology	4 061 282	4 024 965	36 317
海洋生物工程技术 Marine Bioengineering Technology	225 158	214 162	10 996

6-4 续表 continued

行 业 Industry	经费收入总额 Fund Total	经常费 Routine Fund	基本建设中政府投资 Government Investment in the Capital Construction
海洋交通运输工程技术 Marine Communications and Transport Engineering Technology	5 162 549	4 966 188	196 361
海洋能源开发技术 Marine Energies Development Technology	1 050 918	1 050 918	0
海洋环境工程技术 Marine Environmental Engineering Technology	794 903	697 147	97 006
河口水利工程技术 Estuarine Water Conservancy Engineering Technology	2 093 679	2 012 955	80 724
其他海洋工程技术 Other Marine Engineering Technology	536 940	536 940	0
海洋信息服务业 Marine Information Service	**721 727**	**702 939**	**18 788**
其他海洋信息服务 Other Marine Information Services	721 727	702 939	18 788
海洋技术服务业 Marine Technological Service Industry	**1 890 095**	**1 885 011**	**5 084**
海洋工程管理服务 Marine Engineering Management Service	76 130	76 130	0
其他海洋专业技术服务 Other Marine Professional and Technological Services	1 813 965	1 808 881	5 084

6-5 分行业海洋科研机构科技课题情况

Marine Science and Technology Research Projects of the Research Institutions by Industry

单位:项 (item)

行 业 Industry	课题数 Number of Research Projects	基础研究 Basic Research	应用研究 Applied Research	试验发展 Experimental Development	成果应用 Result Application	科技服务 Scientific and Technological Service
合 计 Total	**18 810**	**5 300**	**4 749**	**3 726**	**2 164**	**2 871**
海洋基础科学研究 Marine Basic Scientific Research	**13 868**	**4 969**	**3 998**	**2 078**	**1 095**	**1 728**
海洋自然科学 Marine Natural Science	11 207	4 728	3 620	1 448	526	885
海洋社会科学 Marine Social Science	871	14	20	81	264	492
海洋农业科学 Marine Agricultural Science	1 767	225	351	539	301	351
海洋生物医药 Marine Biomedicine	23	2	7	10	4	0
海洋工程技术研究 Marine Engineering Technology Research	**4 003**	**110**	**438**	**1 543**	**956**	**956**
海洋化学工程技术 Marine Chemical Engineering Technology	1 049	25	122	591	251	60
海洋生物工程技术 Marine Bioengineering Technology	90	0	7	29	45	9

6-5 续表 continued

行 业 Industry	课题数 Number of Research Projects	基础研究 Basic Research	应用研究 Applied Research	试验发展 Experimental Development	成果应用 Result Application	科技服务 Scientific and Technological Service
海洋交通运输工程技术 Marine Communications and Transport Engineering Technology	929	1	29	365	218	316
海洋能源开发技术 Marine Energies Development Technology	324	17	62	139	48	58
海洋环境工程技术 Marine Environmental Engineering Technology	260	6	29	82	43	100
河口水利工程技术 Estuarine Water Conservancy Engineering Technology	1 205	60	165	299	287	394
其他海洋工程技术 Other Marine Engineering Technology	146	1	24	38	64	19
海洋信息服务业 Marine Information Service	**157**	**0**	**2**	**21**	**3**	**131**
其他海洋信息服务 Other Marine Information Services	157	0	2	21	3	131
海洋技术服务业 Marine Technological Service Industry	**782**	**221**	**311**	**84**	**110**	**56**
海洋工程管理服务 Marine Engineering Management Service	49	0	1	12	7	29
其他海洋专业技术服务 Other Marine Professional and Technological Services	733	221	310	72	103	27

6-6 分行业海洋科研机构科技论著情况

Marine Scientific and Technological Works of the Research Institutions by Industry

行 业 Industry	发表科技论文(篇) Scientific Theses Published (piece)	#国外发表 Published Abroad	出版科技著作(种) Scientific and Technological Works Published (kind)
合 计 **Total**	**17 257**	**6 779**	**353**
海洋基础科学研究 **Marine Basic Scientific Research**	**11 184**	**5 062**	**218**
海洋自然科学 Marine Natural Science	8 773	4 600	122
海洋社会科学 Marine Social Science	754	52	64
海洋农业科学 Marine Agricultural Science	1 650	410	32
海洋生物医药 Marine Biomedicine	7	0	0
海洋工程技术研究 **Marine Engineering Technology Research**	**4 364**	**652**	**117**
海洋化学工程技术 Marine Chemical Engineering Technology	887	175	4
海洋生物工程技术 Marine Bioengineering Technology	81	4	1

6-6 续表 continued

行 业 Industry	发表科技论文（篇） Scientific Theses Published (piece)	#国外发表 Published Abroad	出版科技著作(种) Scientific and Technological Works Published (kind)
海洋交通运输工程技术 Marine Communications and Transport Engineering Technology	1 030	166	22
海洋能源开发技术 Marine Energies Development Technology	578	47	4
海洋环境工程技术 Marine Environmental Engineering Technology	186	34	5
河口水利工程技术 Estuarine Water Conservancy Engineering Technology	1 182	190	75
其他海洋工程技术 Other Marine Engineering Technology	420	36	6
海洋信息服务业 Marine Information Service	**297**	**18**	**7**
其他海洋信息服务 Other Marine Information Services	297	18	7
海洋技术服务业 Marine Technological Service Industry	**1 412**	**1 047**	**11**
海洋工程管理服务 Marine Engineering Management Service	79	0	0
其他海洋专业技术服务 Other Marine Professional and Technological Services	1 333	1 047	11

6-7 分行业科研机构科技专利情况
Marine Scientific and Technological Patents of the Research Institutions by Industry

单位:件 (number)

行 业 Industry	专利申请受理数 Number of Patent Applications Accepted	#发明专利 Patents for Discoveries	专利授权数 Number of Patents Granted	#发明专利 Patents for Discoveries	拥有发明专利总数 Total Number of Patents for Discoveries
合 计 Total	**7 176**	**5 824**	**5 622**	**4 111**	**20 518**
海洋基础科学研究 Marine Basic Scientific Research	**2 465**	**1 824**	**1 733**	**1 039**	**4 596**
海洋自然科学 Marine Natural Science	1 884	1 491	1 221	817	3 731
海洋社会科学 Marine Social Science	70	2	10	5	12
海洋农业科学 Marine Agricultural Science	509	329	502	217	853
海洋生物医药 Marine Biomedicine	2	2	0	0	0
海洋工程技术研究 Marine Engineering Technology Research	**3 934**	**3 337**	**3 292**	**2 615**	**12 425**
海洋化学工程技术 Marine Chemical Engineering Technology	2 754	2 651	2 591	2 391	11 227
海洋生物工程技术 Marine Bioengineering Technology	6	5	3	3	6

6-7 续表 continued

行 业 Industry	专利申请受理数 Number of Patent Applications Accepted	#发明专利 Patents for Discoveries	专利授权数 Number of Patents Granted	#发明专利 Patents for Discoveries	拥有发明专利总数 Total Number of Patents for Discoveries
海洋交通运输工程技术 Marine Communications and Transport Engineering Technology	286	114	197	53	261
海洋能源开发技术 Marine Energies Development Technology	379	305	109	47	299
海洋环境工程技术 Marine Environmental Engineering Technology	24	14	19	12	31
河口水利工程技术 Estuarine Water Conservancy Engineering Technology	319	143	274	75	304
其他海洋工程技术 Other Marine Engineering Technology	166	105	99	34	297
海洋信息服务业 Marine Information Service	**0**	**0**	**3**	**2**	**3**
其他海洋信息服务 Other Marine Information Services	0	0	3	2	3
海洋技术服务业 Marine Technological Service Industry	**777**	**663**	**594**	**455**	**3 494**
海洋工程管理服务 Marine Engineering Management Service	4	3	1	1	5
其他海洋专业技术服务 Other Marine Professional and Technological Services	773	660	593	454	3 489

6-8 分行业海洋科研机构R&D情况
R&D in the Marine Scientific Research Institutions by Industry

行 业 Industry	R&D人员 （人） R&D Personnel （Persons）	R&D经费内部支出 （千元） R&D Internal Expenditure （1 000 yuan）	R&D课题数 （项） Number of R&D Projects （item）
合 计 Total	**29 088**	**16 655 582**	**13 775**
海洋基础科学研究 Marine Basic Scientific Research	**18 552**	**11 140 628**	**11 045**
海洋自然科学 Marine Natural Science	16 260	10 098 112	9 796
海洋社会科学 Marine Social Science	291	105 175	115
海洋农业科学 Marine Agricultural Science	1 954	928 895	1 115
海洋生物医药 Marine Biomedicine	47	8 446	19
海洋工程技术研究 Marine Engineering Technology Research	**7 952**	**4 423 600**	**2 091**
海洋化学工程技术 Marine Chemical Engineering Technology	3 100	1 813 909	738
海洋生物工程技术 Marine Bioengineering Technology	190	72 171	36

6-8 续表 continued

行 业 Industry	R&D人员 (人) R&D Personnel (Persons)	R&D经费内部支出 (千元) R&D Internal Expenditure (1 000 yuan)	R&D课题数 (项) Number of R&D Projects (item)
海洋交通运输工程技术 Marine Communications and Transport Engineering Technology	1 876	744 791	395
海洋能源开发技术 Marine Energies Development Technology	1 092	830 532	218
海洋环境工程技术 Marine Environmental Engineering Technology	356	188 193	117
河口水利工程技术 Estuarine Water Conservancy Engineering Technology	1 093	682 220	524
其他海洋工程技术 Other Marine Engineering Technology	245	91 784	63
海洋信息服务业 Marine Information Service	**316**	**133 531**	**23**
其他海洋信息服务 Other Marine Information Services	316	133 531	23
海洋技术服务业 Marine Technological Service Industry	**2 268**	**957 823**	**616**
海洋工程管理服务 Marine Engineering Management Service	260	42 103	13
其他海洋专业技术服务 Other Marine Professional and Technological Services	2 008	915 720	603

6-9 分地区海洋科研机构及人员情况
Marine Scientific Research Institutions and Personnel by Regions

地　区 Region	机构数(个) Number of Institutions (unit)	从业人员(人) Employees (person)
合　计 Total	**192**	**42 331**
北　京 Beijing	23	12 706
天　津 Tianjin	16	2 808
河　北 Hebei	5	552
辽　宁 Liaoning	22	3 151
上　海 Shanghai	15	3 989
江　苏 Jiangsu	10	3 356
浙　江 Zhejiang	21	2 028
福　建 Fujian	14	1 189
山　东 Shandong	22	4 108
广　东 Guangdong	26	5 434
广　西 Guangxi	11	1 225
海　南 Hainan	3	265
其　他 Others	4	1 520

6-10 分地区海洋科研机构科技活动人员学历构成

Educational Background Composition of the Personnel Engaged in Scientific and Technological Activities in Marine Scientific Research Institutions by Regions

单位:人 （person）

地 区 Region	科技活动人员 Personnel Engaged in Scientifical Activities	博士 Doctor	硕士 Master	大学生 Graduate	大专生 College Graduate
合 计 Total	**35 860**	**8 749**	**11 793**	**10 417**	**2 905**
北 京 Beijing	11 569	3 617	3 531	2 847	968
天 津 Tianjin	2 322	218	876	963	134
河 北 Hebei	534	53	134	268	71
辽 宁 Liaoning	2 764	338	1 067	861	271
上 海 Shanghai	3 501	670	1 198	1 145	291
江 苏 Jiangsu	2 068	425	541	626	260
浙 江 Zhejiang	1 723	220	615	687	161
福 建 Fujian	1 129	182	369	443	64
山 东 Shandong	3 279	965	997	819	331
广 东 Guangdong	4 820	1 375	1 771	1 215	221
广 西 Guangxi	661	42	291	256	67
海 南 Hainan	221	17	65	100	23
其 他 Others	1 269	627	338	187	43

6-11 分地区海洋科研机构科技活动人员职称构成

Technical Title Composition of Personel Engaged in Marine Scientific Research Institutions by Regions

单位:人 (person)

地 区 Region	科技活动人员 Personnel Engaged in Scientifical Activities	高级职称 Senior Professional Title	中级职称 Mid-level Professional Title	初级职称 Junior Professional Title
合 计 Total	**35 860**	**14 703**	**11 970**	**5 460**
北 京 Beijing	11 569	5 672	3 727	1 143
天 津 Tianjin	2 322	923	754	400
河 北 Hebei	534	232	144	26
辽 宁 Liaoning	2 764	1 054	898	350
上 海 Shanghai	3 501	1 359	1 304	581
江 苏 Jiangsu	2 068	890	587	350
浙 江 Zhejiang	1 723	657	553	266
福 建 Fujian	1 129	381	433	234
山 东 Shandong	3 279	1 224	1 285	553
广 东 Guangdong	4 820	1 481	1 437	1 195
广 西 Guangxi	661	218	261	164
海 南 Hainan	221	46	50	86
其 他 Others	1269	566	537	112

6-12 分地区海洋科研机构经费收入
Routine Fund Receipts of Marine Scientific Research Institutions by Regions

单位:千元 (1 000 yuan)

地区 Region	经费收入总额 Fund Total	本年收入合计 Total Annual Income	基本建设中政府投资 Government Investment in the Capital Construction
合计 Total	**33 334 007**	**31 020 439**	**2 312 818**
北京 Beijing	10 378 485	9 457 253	921 232
天津 Tianjin	1 787 632	1 644 860	142 772
河北 Hebei	187 038	173 248	13 790
辽宁 Liaoning	2 298 284	2 278 184	20 100
上海 Shanghai	3 730 432	3 540 832	189 600
江苏 Jiangsu	2 645 398	2 465 023	180 375
浙江 Zhejiang	1 676 635	1 541 770	134 115
福建 Fujian	824 865	741 703	83 162
山东 Shandong	3 774 704	3 412 012	362 692
广东 Guangdong	4 134 619	3 884 425	250 194
广西 Guangxi	861 614	857 394	4 220
海南 Hainan	97 201	95 255	1 946
其他 Others	937 100	928 480	8 620

6-13 分地区海洋科研机构科技课题情况
Marine Science and Technology Research Projects of the Research Institutions by Regions

单位：项　　　　(item)

地 区 Region	课题数 Number of Research Projects	基础研究 Basic Research	应用研究 Applied Research	试验发展 Experimental Development	成果应用 Result Application	科技服务 Scientific and Technological Service
合 计 Total	**18 810**	**5 300**	**4 749**	**3 726**	**2 164**	**2 871**
北 京 Beijing	6 741	1 990	1 657	908	773	1 413
天 津 Tianjin	790	2	67	373	88	260
河 北 Hebei	77	14	24	23	8	8
辽 宁 Liaoning	685	47	207	224	175	32
上 海 Shanghai	1 166	119	317	336	165	229
江 苏 Jiangsu	2 187	366	678	514	447	182
浙 江 Zhejiang	645	127	44	129	123	222
福 建 Fujian	618	193	245	92	41	47
山 东 Shandong	1 637	477	556	257	226	121
广 东 Guangdong	2 653	957	728	576	84	308
广 西 Guangxi	125	21	28	58	11	7
海 南 Hainan	25	0	8	5	10	2
其 他 Others	1 461	987	190	231	13	40

6-14 分地区海洋科研机构科技论著情况
Marine Scientific and Technological Works of the Research Institutions by Regions

地 区 Region	发表科技论文(篇) Scientific Theses Published (piece)	#国外发表 Published Abroad	出版科技著作(种) Scientific and Technological Works Published (kind)
合 计 Total	**17 257**	**6 779**	**353**
北 京 Beijing	5 657	2 146	129
天 津 Tianjin	886	200	22
河 北 Hebei	421	2	45
辽 宁 Liaoning	775	296	14
上 海 Shanghai	826	158	7
江 苏 Jiangsu	1 132	407	21
浙 江 Zhejiang	554	109	15
福 建 Fujian	381	175	6
山 东 Shandong	2 009	834	45
广 东 Guangdong	2 929	1 626	39
广 西 Guangxi	167	21	0
海 南 Hainan	97	9	1
其 他 Others	1 423	796	9

6-15 分地区海洋科研机构科技专利情况
Marine Scientific and Technological Patents of the Research Institutions by Regions

单位:件 （number）

地 区 Region	专利申请受理数 Number of Patent Applications Accepted	#发明专利 Patents for Discoveries	专利授权数 Number of Patents Granted	#发明专利 Patents for Discoveries	拥有发明专利总数 Total Number of Patents for Discoveries
合 计 Total	**7 176**	**5 824**	**5 622**	**4 111**	**20 518**
北 京 Beijing	2 647	2 282	2 004	1 681	8 330
天 津 Tianjin	188	78	121	42	199
河 北 Hebei	4	0	4	0	0
辽 宁 Liaoning	878	796	816	711	3 093
上 海 Shanghai	1 170	963	836	600	2 882
江 苏 Jiangsu	320	202	181	81	349
浙 江 Zhejiang	154	112	158	73	202
福 建 Fujian	55	50	48	43	106
山 东 Shandong	561	408	505	254	1 136
广 东 Guangdong	976	797	772	546	3 771
广 西 Guangxi	48	25	47	13	74
海 南 Hainan	0	0	8	8	8
其 他 Others	175	111	122	59	368

6-16 分地区海洋科研机构R&D情况

R&D in the Marine Scientific Research Institutions by Regions

行 业 Industry	R&D人员 (人) R&D Personnel (persons)	R&D经费内部支出 (千元) R&D Internal Expenditure (1 000 yuan)	R&D课题数 (项) Number of R&D Projects (item)
合 计 Total	**29 088**	**16 655 582**	**13 775**
北 京 Beijing	9 527	6 092 467	4 555
天 津 Tianjin	1 775	895 039	442
河 北 Hebei	293	91 209	61
辽 宁 Liaoning	1 659	1 023 051	478
上 海 Shanghai	3 067	2 307 215	772
江 苏 Jiangsu	1 480	830 532	1 558
浙 江 Zhejiang	818	556 497	300
福 建 Fujian	809	366 234	530
山 东 Shandong	2 471	1 801 784	1 290
广 东 Guangdong	4 465	1 778 965	2 261
广 西 Guangxi	413	84 997	107
海 南 Hainan	26	3 001	13
其 他 Others	2 285	824 591	1 408

主要统计指标解释

1. 海洋科研机构 指有明确的研究方向和任务，有一定水平的学术带头人和一定数量、质量的研究人员，有开展研究工作的基本条件，长期有组织地从事海洋研究与开发活动的机构。

2. 从业人员 指由本机构年末直接组织安排工作并支付工资的各类人员总数。包括固定职工、国家有编制的合同制职工、招聘人员和返聘的离退休人员。不包括离退休人员、停薪留职人员。

3. 从事科技活动人员 指从业人员中的科技管理人员、课题活动人员和科技服务人员。

4. 高级职称 指研究员、副研究员；教授、副教授；高级工程师；高级农艺师；正、副主任医（药、护、技）师；高级实验师；高级统计师；高级经济师；高级会计师；编审（正、副编审）；译审（正、副译审）、高级（主任）记者；正、副研究馆员等。

5. 中级职称 指助理研究员；讲师；工程师；农艺师；主治医（药、护、技）师；实验师；统计师；经济师；会计师；编辑；翻译；记者；馆员等。

6. 初级职称 指研究实习员；助教；助理工程师、技术员；助理农艺师、农业技术员；医（药、护、技）师、医（药、护、技）士；助理实验师、实验员；助理统计师、统计员；助理经济师；助理会计师、会计员；助理编辑、见习编辑；助理翻译；助理记者；助理馆员、管理员等。

7. 科技经费筹集额 指从各种渠道筹集到的计划用于本单位科技活动的经费，不论来源渠道如何。

8. 政府资金 指由各级政府部门直接拨款或企事业单位利用政府资金委托本机构从事科学技术活动所获得的收入。

9. 生产经营活动收入 指本机构在科研、技术等专业业务活动以外开展非独立核算的经营活动取得的收入，包括产品（商品）销售收入、经营服务收入、工程承包收入、租赁收入和其他经营收入。

10. 其他收入 指开展科技活动与生产经营活动以外的各项活动的收入，包括：用于离退休人员的政府拨款。

11. 非科技活动借贷款 指本机构为开展非科技活动从各种渠道获得的各类借、贷款。不论偿还形式、期限和数额如何，均按当年获得的借、贷款额填报。不包括基本建设贷款。

12. 基础研究 为获得新知识而进行的独创性研究。其目的是揭示观察到的现象和事实的基本原理和规律，而不以任何特定的实际应用为目的。

13. 应用研究 为获得新的科学技术知识而进行的独创性研究。它主要针对某一特定的实际应用目的。应用研究通常是为了确定基础研究成果或知识的可能的用途，或是为达到某一具体的、预定的实际目的确定新的方法（原理性）或途径。

14. 试验发展 利用从研究或实际经验获得的知识，为生产新的材料、产品和装置，建立新的工艺和系统，以及对已生产或建立的上述各项进行实质性的改进而进行的系统性工作。

15. 成果应用 为解决R&D活动阶段产生的新产品、新装置、新工艺、新技术、新方法、新系统和服务等能投入生产或在实际应用中所存在的技术问题而进行的系统性活动。它不具有创新成分。此类活动包括为达到生产目的而进行的定型设计和试制以及为扩大新产品的生产规模和探索新方法、新技术、新工艺等的应用领域而进行的适应性试验。

16. 科技服务 与科学研究与实验发展有关,并有助于科学技术知识的产生、传播和应用的活动。包括为扩大科技成果的使用范围而进行的示范性推广工作;为用户提供科技信息和文献服务的系统性工作;为用户提供可行性报告、技术方案、建议及进行技术论证等技术咨询工作;自然、生物现象的日常观测、监测,资源的考查和勘探;有关社会、人文、经济现象的通用资料的收集,如统计、市场调查等以及这些资料的常规分析与整理;为社会和公众提供的测试、标准化、计量、计算、质量控制和专利服务,不包括工商企业为进行正常生产而开展的上述活动。

17. 生产性活动 由于业务特殊的工艺设备条件,或掌握某种技术专长或诀窍,所进行的小量非常规生产。

18. 科技论文 在全国性学报或学术刊物上、省部属大专院校对外正式发行的学报或学术刊物上发表的论文以及向国外发表的论文。

19. 科技著作 经过正式出版部门编印出版的科技专著、大专院校教科书、科普著作。

20. 专利申请受理数 当年本单位向专利管理部门提出申请并被受理的职务专利申请件数。

21. 专利授权数 当年由专利管理部门授予本单位专利权的职务专利件数。

Explanatory Notes on Main Statistical Indicators

1. Marine Scientific Research Institution refers to the institution which has definite research orientations and tasks, high-level academic leading personnel and fair-sized, qualified research personnel, and basic conditions for research work and which is engaged for a long time in the marine research and development activities in an organized way.

2. Employees refer to the total number of personnel of various kinds employed and paid by the institution at the end of the year, including fixed employees, contract workers of staff belonging to the state authorized staff , recruited personnel, reemployed retired personnel, but not including the retired and the personnel on leave with pay suspension.

3. Personnel Engaged in Scientific and Technological Activities refers to the personnel for scientific and technological management, personnel engaged in the activities of research topics and scientific and technological service personnel.

4. Senior Professional Title refer to research scientist, associate research scientist; professor, associate professor; senior engineer; senior agronomist; professor-rank and associate professor-rank doctor (pharmacists, nurses and technicians); senior laboratory technician; senior statisticians; senior economic engineer; chief accountant; senior editor (professor and associate professor ranks); senior translator (professor and associate professor ranks); senior journalist; research librarian (professor and associate professor ranks), etc.

5. Mid-level Professional Title refer to assistant research scientist; lecturer; engineer; agronomist; lecturer-rank doctor (pharmacist, nurse and technician); laboratory technician; lecturer-rank statistician; economic engineer; accountant; editor; translator; journalist; librarian, etc.

6. Junior Professional Title refers to trainee researcher; assistant; assistant engineer, technician, assistant agronomist, agricultura technician; assistant-rank doctor (pharmacist, nurse and technician); assistant laboratory technician; assistant statistician; assistant economic engineer, assistant accountant; assistant editor, editor on probation; assistant translator; assistant journalist; assistant research librarian, librarian, etc.

7. Amount of Scientific and Technological Funds Raised refers to the funds raised through all channels planned to be used as funds for the scientific and technological activities in the institution regardless of their source and channels.

8. Funds from Government refers to the direct appropriations by the government departments at all levels or the earnings from conducting scientific and technological activities entrusted to the institution by enterprises as institutions by earning the funds from government.

9. Earnings from Production as Business Activities refer to the incomes obtained from the non-independent accounting business activities carried out by the institution beyond the scientific research, technological and professional activities, including these from sale of products(goods), business and service, contracted projects, leasing and other business.

10. Other Incomes refer to those from the various activities conducted other than scientific and technological activities and production and business activities, including the government appropriations for retired personnel.

11. Loan for Non-scientific as Technological Activities refers to the various types loan obtained by the institutions through all channels for carrying out non-scientific and technological activities, not including the load for capital construction. The loan, irrespective of its form of reimbursement, term and amount is filled in a form and submitted to the authorities as the amount acquired in the current year.

12. Basic Research refers to the original research to acquire new knowledge. It is aimed at revealing the basic principles and laws of the phenomena and facts observed, but not at any specific practical applications.

13. Applied Research refers to the original research to acquire new scientific and technological knowledge. It mainly serves the purpose of a particular practical application. The purpose of applied research is usually to define the potential uses of the research finds or knowledge obtained from basic research or to identify new methods (principles) or ways to reach a specific and predetermined goal.

14. Experimental Development refers to the systematic work carried out to establish new technologies and systems for producing new materials, products and equipment by using the knowledge obtained from research or practical experience, or to make substantial improvement of the above-mentioned which have been produced or established.

15. Result Application refers to the systematic activities carried out to solve the technical problems that might crop up in the production or practical application of the new products, devices, technologies, techniques, methods, systems and service occurring in the course of R & D activities. They do not bring forth new ideas. Such activities include the finalizing design and trial-production for the purpose of production as well as the adaptive tests to expand the production scale of new products and the application areas of new methods, techniques and technologies.

16. Scientific and Technological Service refers to the activities that are associated with the scientific research and experimental development, and contribute to the generation, dissemination and application of scientific and technological knowledge, which include the demonstrative work of popularization to enlarge the use scope of scientific and technological achievements; the systematic work of providing the users with scientific and technological information and literature service; the technical consultation work of providing users with feasibility reports, technical schemes and recommendations and carrying out technical demonstration; routine observation and monitoring of natural and biological phenomena, and the survey and exploration of resources; collection of universal data on the appropriate social, cultural and economic phenomena, such as statistics and market survey, as well as the routine analysis and sorting-out of these data; the provision for the society and the public of such service as testing, standardization, computation, quality control and patent, but not including the type of the above-mentioned activities carried out by industrial and commercial enterprises for the purpose of normal production.

17. Productive Activity refers to the small-scale and non-conventional productions due to the presence of special technologies and equipment or mastery of a particular technical expertise or secret of success.

18. Scientific Treatises refer to the theses published in the national journals or academic publications, those officially issued journals or academic publications by universities and colleges under provinces or ministries as well as theses published abroad.

19. Scientific and Technological Works refer to the scientific and technological monographs, textbooks for universities and colleges and popular science books published by the official publishing houses.

20. Number of Patents Applied and Accepted refers to the number of professional patent applications of the unit to the patent administrative department and accepted by it in the year.

21. Number of Patents Granted refers to the number of the professional patents granted to the unit by the patent administrative department in the year.

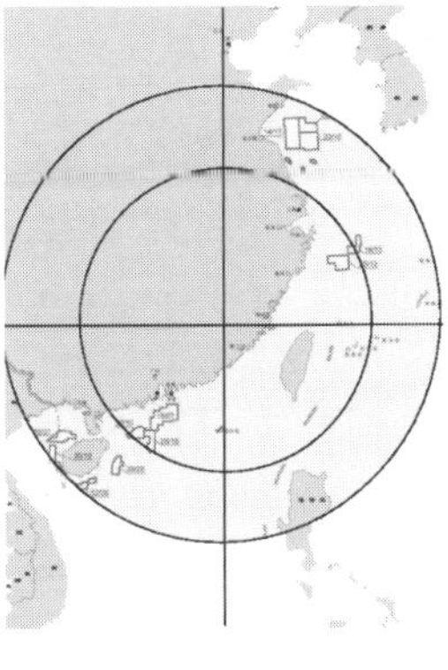

7
海 洋 教 育
Marine Education

7-1 全国各海洋专业博士研究生情况
Doctoral Students from Marine Specialities

专 业 Speciality	专业点数(个) Number of Speciality Agencies	学生数(人) Number of Students (person)			
		毕业生 Graduates	招 生 Entrants	在校生 Enrollment	预计毕业生数 Estimated Number of Graduates
合 计 Total	**140**	**630**	**979**	**4 313**	**2 111**
物理海洋学 Physical Oceanography	5	20	39	156	74
海洋化学 Marine Chemistry	4	17	30	107	46
海洋生物学 Marine Biology	8	54	82	283	103
海洋地质 Marine Geology	7	35	43	152	70
海洋科学学科 Marine Science Subjects	7	91	150	568	280
水生生物学 Hydrobiology	19	85	76	330	175
水文学及水资源 Hydrology and Water Resource	19	87	128	673	382
港口海岸及近海工程 Coastal Harbour and Offshore Engineering	10	37	62	332	168

7-1 续表 continued

专 业 Speciality	专业点数(个) Number of Speciality Agencies	学生数(人) Number of Students (person)			
		毕业生 Graduates	招 生 Entrants	在校生 Enrollment	预计毕业生数 Estimated Number of Graduates
船舶与海洋结构物设计制造 Ships and Marine Structures Design and Manufacture	12	41	59	383	244
轮机工程 Turbine Engineering	8	38	52	286	109
水声工程 Hydroacoustic Engineering	7	38	37	189	91
船舶与海洋工程学科 Ship and Ocean Engineering Subjects	7	13	79	282	87
水产品加工及贮藏工程 Aquatic Products Processing and Storing Engineering	7	8	2	59	51
水产养殖 Aquaculture	7	30	52	213	106
捕捞学 Science of Fishing	2	11	21	68	30
渔业资源 Fishery Resource	4	7	21	76	36
水产学科 Fishery Subjects	6	18	46	155	58
航空、航天与航海医学 Aeronautical, Aerospace and Nautical Medicine	1	0	0	1	1

7-2 全国各海洋专业硕士研究生情况
Postgraduate Students from Marine Specialities

专 业 Speciality	专业点数(个) Number of Speciality Agencies	学生数(人) Number of Students (person)			
		毕业生 Graduates	招 生 Entrants	在校生 Enrollment	预计毕业生数 Estimated Number of Graduates
合 计 Total	**322**	**2 961**	**3 357**	**10 215**	**3 502**
物理海洋学 Physical Oceanography	14	66	162	447	143
海洋化学 Marine Chemistry	16	102	106	349	128
海洋生物学 Marine Biology	23	241	274	837	271
海洋地质 Marine Geology	15	127	153	469	159
海洋科学学科 Marine Science Subjects	13	182	319	890	250
水生生物学 Hydrobiology	44	200	216	659	237
水文学及水资源 Hydrology and Water Resource	46	344	383	1 179	381
港口海岸及近海工程 Coastal Harbour and Offshore Engineering	22	270	260	781	287

7-2 续表 continued

专 业 Speciality	专业点数(个) Number of Speciality Agencies	学生数(人) Number of Students (person)			
		毕业生 Graduates	招 生 Entrants	在校生 Enrollment	预计毕业生数 Estimated Number of Graduates
船舶与海洋结构物设计制造 Ships and Marine Structures Design and Manufacture	20	328	290	884	329
轮机工程 Turbine Engineering	12	213	243	707	248
水声工程 Hydroacoustic Engineering	9	111	110	359	122
船舶与海洋工程学科 Ship and Ocean Engineering Subjects	14	166	228	703	234
水产品加工及贮藏工程 Aquatic Products Processing and Storing Engineering	25	103	56	216	96
水产养殖 Aquaculture	28	347	354	1 170	415
捕捞学 Science of Fishing	4	25	16	61	27
渔业资源 Fishery Resource	9	73	74	232	87
水产学科 Fishery Subjects	7	49	113	272	88
航空、航天与航海医学 Aeronautical, Aerospace and Nautical Medicine	1	14	0	0	0

7-3 全国普通高等教育各海洋专业本科学生情况
Undergraduates from Marine Specialities in the National Ordinary Higher Education

专 业 Speciality	专业点数（个）Number of Speciality Agencies	学生数（人）Number of Students（person）			
		毕业生 Graduates	招 生 Entrants	在校生 Enrollment	预计毕业生数 Estimated Number of Graduates
合 计 Total	**268**	**15 026**	**18 202**	**69 233**	**16 663**
海洋科学 Marine Science	25	862	1 005	4 221	953
海洋技术（注：可授理学或工学学士学位）Marine Technology（Note: It may confer bachelor′s degrees in science and engineering）	22	619	771	2 814	662
海洋资源与环境 Marine Living Resources and Environment	11	175	370	1 380	260
海洋科学类专业 New Specialties Under the Category of Marine Sciences	9	98	777	1 313	105
港口航道与海岸工程 Harbour Channel and Coastal Engineering	30	1 929	1 552	7 357	2 053
航海技术 Nautical Technology	17	2 193	2 581	10 177	2 488
轮机工程 Turbine Engineering	23	2 781	2 964	12 385	3 097
船舶与海洋工程 Ship and Marine Engineering	34	2 924	2 771	10 667	2 907
海洋工程与技术 Ocean Engineering and Technology	5	67	134	438	132
海洋资源开发技术 Marine Resources Exploitation Technology	9	77	330	1 055	204
海洋工程类专业 New Speciality of Marine Engineering	3	0	274	580	0
水产养殖学 Aquaculture	55	2 540	3 127	12 530	2 968
海洋渔业科学与技术 Science and Technology of Marine Fishery	10	393	695	2 127	460
水族科学与技术 Science and Technology of Aquatic Animals	10	254	415	1 378	263
水产类专业 Fishery-type Specialities	2	0	255	319	0
海事管理 Maritime Affairs Management	3	114	181	492	111

7-4 全国普通高等教育各海洋专业专科学生情况

Students from Marine Specialities of the Colleges for Professional Training in the National Ordinary Higher Education

专业 Speciality	专业点数（个） Number of Speciality Agencies	学生数(人) Number of Students (person)			
		毕业生 Graduates	招生 Entrants	在校生 Enrollment	预计毕业生数 Estimated Number of Graduates
合计 Total	**506**	**30 318**	**24 803**	**82 165**	**29 259**
水产养殖技术 Aquaculture Technology	38	1 234	1 162	3 337	1 021
渔业综合技术 Integrated Fishery Technology	2	7	39	47	8
水运管理 Water Transport Management	2	62	0	47	47
国际航运业务管理 International Shipping Business Management	33	2 179	1 905	6 431	2 230
轮机工程技术 Turbine Engineering Technology	51	4 876	3 923	12 556	4 440
船舶工程技术 Ship Engineering Technology	43	4 543	3 322	11 329	4 023
船舶检验 Ship Inspection	8	70	210	460	84
船机制造与维修 Ship Engines Manufacturing and Maintenance	7	385	411	1 194	368
船舶舾装 Ship Equipment and Installations	4	102	141	331	99
港口业务管理 Harbour Business Management	19	905	806	2 751	863
港口物流设备与自动控制 Harbour Logistics Equipment and Automatic Control	19	1 008	703	2 512	905

7-4 续表 continued

专 业 Speciality	专业点数（个） Number of Speciality Agencies	学生数(人) Number of Students (person)			
		毕业生 Graduates	招 生 Entrants	在校生 Enrollment	预计毕业生数 Estimated Number of Graduates
集装箱运输管理 Container Transportation Management	16	688	523	1 761	681
港口工程技术 Harbour Engineering	11	595	350	1 173	474
报关与国际货运 Customs Clearing and International Freight Transport	212	12 513	9 494	33 200	12 482
港口与航运管理 Harbour and Shipping Management	10	456	648	1 953	575
港口机械应用技术 Applied Harbour Machinery Technology	6	283	359	798	190
港口运输类专业 Majors of Port Transportation	4	0	238	271	33
水文自动化测报技术 Automatic Hydrological Observing Technology	1	17	0	28	28
水信息技术 Hydrological Information Technology	1	8	6	11	0
水政水资源管理 Water Resources Management by Water Administration	4	99	313	694	222
港口航道与治河工程 Harbour, Channel and River Harnessing Engineering	4	94	53	304	118
水环境监测与保护 Water Environmental Monitoring and Protection	6	145	175	750	289
海关管理 Customs Management	5	49	22	227	79

7-5 全国成人高等教育各海洋专业本科学生情况
Undergraduates from Marine Specialities in the National Adult Higher Education

专业 Speciality	专业点数（个） Number of Speciality Agencies	学生数（人）Number of Students (person)			
		毕业生 Graduates	招生 Entrants	在校生 Enrollment	预计毕业生数 Estimated Number of Graduates
合计 Total	**61**	**1 858**	**2 068**	**5 298**	**2 347**
海洋科学 Marine Science	1	0	10	19	0
海洋技术 （注：可授理学或工学学士学位） Marine Technology (Note: It may confer bachelor's degrees in science and engineering)	2	37	0	118	91
港口航道与海岸工程 Harbour Channel and Coastal Engineering	3	26	38	72	19
航海技术 Navigation Technology	8	198	224	597	237
轮机工程 Turbine Engineering	11	263	276	866	445
船舶与海洋工程 Ship and Marine Engineering	13	1 167	1 236	2 907	1 301
海洋渔业科学与技术 Science and Technology of Marine Fishery	1	0	1	1	0
水产养殖学 Aquaculture	21	167	283	718	254
海事管理 Maritime Affairs Management	1	0	0	0	0

7-6 全国成人高等教育各海洋专业专科学生情况

Students from Marine Specialities of the Colleges for Professional Training in the National Adult Higher Education

专业 Speciality	专业点数（个） Number of Speciality Agencies	学生数(人) Number of Students (person)			
		毕业生 Graduates	招生 Entrants	在校生 Enrollment	预计毕业生数 Estimated Number of Graduates
合计 Total	**146**	**5 968**	**2 503**	**8 632**	**4 261**
水产养殖技术 Aquaculture Technology	22	94	457	1 450	259
渔业综合技术 Ingrated Fishery Technology	1	0	0	1	0
水产养殖类新专业 New Specialties under the Category of Aquaculture	2	8	5	16	5
国际航运业务管理 International Shipping Business Management	5	182	60	203	71
轮机工程技术 Turbine Engineering Technology	33	2 988	668	3 128	1 758
船舶工程技术 Ship Engineering	24	1 657	995	2 528	1 373
船舶检验 Ship Inspection	2	6	2	24	3
船机制造与维修 Ship Engines Manufacturing and Maintenance	2	96	2	35	20
水上运输类新专业 New Specialties under the Category of Water Transport	3	36	19	55	36

7-6 续表 continued

专 业 Speciality	专业点数（个）Number of Speciality Agencies	学生数(人) Number of Students (person)			
		毕业生 Graduates	招 生 Entrants	在校生 Enrollment	预计毕业生数 Estimated Number of Graduates
港口业务管理 Harbour Business Management	3	0	3	4	0
港口物流设备与自动控制 Harbour Logistics Equipment and Automatic Control	1	50	55	154	48
报关与国际货运 Customs Declaration/International Freight Transport	38	824	179	784	526
港口与航运管理 Harbour and Shipping Management	3	1	16	161	145
水文自动化测报技术 Automatic Hydrological Observing Technology	1	0	0	1	0
水政水资源管理 Water Resources Management by Water Administration	4	26	34	62	17
港口航道与治河工程 Harbour, Channel and River Harnessing Engineering	1	0	0	10	0
海关管理 Customs Management	1	0	8	16	0

7-7 全国中等职业教育各海洋专业学生情况
Students from Marine Specialities in the National Secondary Vocational Education

专 业 Speciality	专业点数（个） Number of Speciality Agencies	学生数(人) Number of Students (person)			
		毕业生 Graduates	招 生 Entrants	在校生 Enrollment	预计毕业生数 Estimated Number of Graduates
合 计 Total	**337**	**18 250**	**15 739**	**39 472**	**13 778**
海水生态养殖 Seawater Ecological Cultivation	15	1 024	563	1 586	416
航海捕捞 Sea Fishing	2	36	214	490	114
农林牧渔类新专业 New Specialities of Agriculture, Forestry, Animal Husbandry and Fishery	39	3 671	2 416	6 654	2 976
水文与水资源勘测 Hydrological and Water Resources Survey	5	70	46	257	104
风电场机电设备运行与维护 Operation and Maintenance of Electromechanical Equipment in the Wind Power Station	34	1 350	1 420	3 564	1 087
船舶制造与修理 Ships Building and Repair	45	1 956	1 991	7 118	2 609
船舶机械装置安装与维修 Installation and Maintenance of Ships' Mechanical Equipment	8	541	607	1 355	412
船舶驾驶 Ship Piloting	64	4 723	4 225	7 776	2 518

7-7 续表 continued

专 业 Speciality	专业点数（个） Number of Speciality Agencies	学生数(人) Number of Students (person)			
		毕业生 Graduates	招 生 Entrants	在校生 Enrollment	预计毕业生数 Estimated Number of Graduates
轮机管理 Engines Management	54	2 858	2 829	5 162	1 632
船舶水手与机工 Ship Sailors and Mechanics	16	312	211	1 242	478
船舶电气技术 Ship Electric Technology	17	324	288	1 179	372
船舶通信与导航 Marine Communication and Navigation	1	0	0	2	2
外轮理货 Foreign Ships Freight Forwarding	10	317	137	603	231
船舶检验 Ships Inspection	3	47	0	42	12
港口机械运行与维护 Operation and Maintenance of Harbour Machinery	22	972	752	2 371	815
工程潜水 Engineering Diving	2	49	40	71	0

注:此表不包含技工学校相关数据。

Note: Data related to Technical Schools are not included in the table.

7-8 分地区各海洋专业博士研究生情况
Doctoral Students in the Marine Specialities by Regions

地 区 Region	专业点数(个) Number of Speciality Agencies	学生数(人) Number of Students (person)			
		毕业生 Graduates	招 生 Entrants	在校生 Enrollment	预计毕业生数 Estimated Number of Graduates
合 计 Total	**140**	**630**	**979**	**4 313**	**2 111**
北 京 Beijing	13	160	207	684	263
天 津 Tianjin	5	4	7	28	8
辽 宁 Liaoning	8	30	47	340	183
上 海 Shanghai	16	51	99	441	197
江 苏 Jiangsu	11	35	90	539	296
浙 江 Zhejiang	7	7	39	150	61
福 建 Fujian	8	28	63	208	88
山 东 Shandong	13	124	180	748	396
广 东 Guangdong	10	23	37	140	75
广 西 Guangxi	1	0	2	4	1
海 南 Hainan	1	0	2	7	3
其 他 Others	47	168	206	1 024	540

7-9 分地区各海洋专业硕士研究生情况
Postgraduate Students in the Marine Specialities by Regions

地区 Region	专业点数(个) Number of Speciality Agencies	学生数(人) Number of Students (person)			
		毕业生 Graduates	招生 Entrants	在校生 Enrollment	预计毕业生数 Estimated Number of Graduates
合计 Total	**322**	**2 961**	**3 357**	**10 215**	**3 502**
北京 Beijing	23	177	270	787	238
天津 Tianjin	9	65	72	241	86
河北 Hebei	4	4	11	36	12
辽宁 Liaoning	23	370	338	1 050	380
上海 Shanghai	32	355	372	1 128	428
江苏 Jiangsu	25	344	364	1 087	365
浙江 Zhejiang	25	172	292	787	224
福建 Fujian	21	158	177	549	194
山东 Shandong	32	331	370	1 151	387
广东 Guangdong	17	188	148	502	193
广西 Guangxi	3	15	16	44	15
海南 Hainan	4	19	21	67	26
其他 Others	104	763	906	2 786	954

7-10 分地区普通高等教育各海洋专业本科学生情况
Undergraduates in the Marine Specialities of Ordinary Higher Education by Regions

地 区 Region	专业点数(个) Number of Speciality Agencies	学生数(人) Number of Students (person)			
		毕业生 Graduates	招 生 Entrants	在校生 Enrollment	预计毕业生数 Estimated Number of Graduates
合 计 Total	**268**	**15 026**	**18 202**	**69 233**	**16 663**
北 京 Beijing	3	113	119	386	88
天 津 Tianjin	13	823	863	3 345	922
河 北 Hebei	10	168	510	1 617	258
辽 宁 Liaoning	24	2 146	2 212	8 803	2 196
上 海 Shanghai	18	1 205	1 374	5 089	1 226
江 苏 Jiangsu	34	1 433	1 596	5 795	1 531
浙 江 Zhejiang	32	898	1 277	4 535	1 059
福 建 Fujian	15	1 148	1 614	6 218	1 658
山 东 Shandong	32	1 832	2 352	8 584	2 137
广 东 Guangdong	19	838	1 529	6 217	1 022
广 西 Guangxi	7	175	390	1 153	237
海 南 Hainan	4	114	344	762	147
其 他 Others	57	4 133	4 022	16 729	4 182

7-11 分地区普通高等教育各海洋专业专科学生情况

Students from of the Marine Specialities Colleges for Professional Training in the Ordinary Higher Education by Regions

地 区 Region	专业点数(个) Number of Speciality Agencies	学生数(人) Number of Students (person)			
		毕业生 Graduates	招 生 Entrants	在校生 Enrollment	预计毕业生数 Estimated Number of Graduates
合 计 Total	**506**	**30 318**	**24 803**	**82 165**	**29 259**
北 京 Beijing	1	0	11	70	37
天 津 Tianjin	12	1 284	1 232	3 565	1 164
河 北 Hebei	31	633	450	1 562	573
辽 宁 Liaoning	35	3 610	2 967	9 167	3 261
上 海 Shanghai	29	1 808	1 139	4 254	1 664
江 苏 Jiangsu	58	4 139	2 736	10 217	3 947
浙 江 Zhejiang	30	1 987	1 583	5 546	2 065
福 建 Fujian	32	1 517	1 367	4 476	1 608
山 东 Shandong	70	4 867	4 477	13 854	4 583
广 东 Guangdong	33	1 814	1 727	5 905	2 295
广 西 Guangxi	28	1 139	1 005	3 368	1 123
海 南 Hainan	6	526	149	928	344
其 他 Others	141	6 994	5 960	19 253	6 595

7-12 分地区成人高等教育各海洋专业本科学生情况
Students from Marine Specialities in the Adult Higher Education by Regions

地 区 Region	专业点数(个) Number of Speciality Agencies	学生数(人) Number of Students (person)			
		毕业生 Graduates	招 生 Entrants	在校生 Enrollment	预计毕业生数 Estimated Number of Graduates
合 计 Total	**60**	**1 858**	**2 068**	**5 298**	**2 347**
天 津 Tianjin	2	46	23	104	52
河 北 Hebei	1	9	8	23	11
辽 宁 Liaoning	5	287	229	479	250
上 海 Shanghai	4	15	159	582	262
江 苏 Jiangsu	4	831	991	2 251	1 098
浙 江 Zhejiang	6	96	85	300	99
福 建 Fujian	1	0	0	18	18
山 东 Shandong	10	63	111	220	53
广 东 Guangdong	5	162	103	359	169
广 西 Guangxi	2	0	3	4	1
其 他 Others	20	349	356	958	334

7-13 分地区成人高等教育各海洋专业专科学生情况
Students from Marine Specialities of the Colleges for Professional Training in the Adult Higher Education by Regions

地 区 Region	专业点数(个) Number of Speciality Agencies	学生数(人) Number of Students (person)			
		毕业生 Graduates	招 生 Entrants	在校生 Enrollment	预计毕业生数 Estimated Number of Graduates
合 计 Total	**146**	**5 968**	**2 503**	**8 632**	**4 261**
天 津 Tianjin	3	167	24	104	54
河 北 Hebei	1	0	1	5	3
辽 宁 Liaoning	20	1 813	537	1 823	1 216
上 海 Shanghai	11	413	83	692	285
江 苏 Jiangsu	20	1 457	765	2 111	1 150
浙 江 Zhejiang	8	189	97	438	337
福 建 Fujian	5	133	318	936	20
山 东 Shandong	18	320	96	313	120
广 东 Guangdong	15	424	119	455	216
广 西 Guangxi	4	0	16	30	14
海 南 Hainan	2	0	3	3	0
其 他 Others	39	1 052	444	1 722	846

7-14 分地区中等职业教育各海洋专业学生情况
Students from Marine Specialities in the Secondary Vocational Education by Regions

地 区 Region	专业点数(个) Number of Speciality Agencies	学生数(人) Number of Students (person)			
		毕业生 Graduates	招 生 Entrants	在校生 Enrollment	预计毕业生数 Estimated Number of Graduates
合 计 Total	**337**	**18 250**	**15 739**	**39 472**	**13 778**
天 津 Tianjin	5	549	709	1 588	595
河 北 Hebei	21	1 476	710	2 398	764
辽 宁 Liaoning	31	1 966	550	1 862	867
上 海 Shanghai	15	1 294	1 325	3 191	967
江 苏 Jiangsu	38	899	651	2 191	825
浙 江 Zhejiang	18	923	1120	2 884	1 269
福 建 Fujian	33	2 733	4 509	6 646	1 503
山 东 Shandong	49	3 431	1 561	4 254	1 696
广 东 Guangdong	10	375	343	1 007	272
广 西 Guangxi	10	480	890	2 672	869
海 南 Hainan	5	80	71	184	57
其 他 Others	102	4 044	3 300	10 595	4 094

7-15 分地区开设海洋专业高等学校教职工数
Number of Teaching and Administrative Staff in the Universities and Colleges Offering Marine Specialities by Regions

地 区 Region	学校(机构)数(个) Number of Colleges (Institutions)	教职工数(人) Number of Teaching and Administrative Staff (person)	专任教师数(人) Number of Full-Time Teachers (person)
合 计 Total	**415**	**497 983**	**322 942**
北 京 Beijing	4	7 231	4 319
天 津 Tianjin	12	16 077	11 222
河 北 Hebei	25	22 711	14 509
辽 宁 Liaoning	23	20 716	13 720
上 海 Shanghai	17	21 301	10 735
江 苏 Jiangsu	47	64 642	42 115
浙 江 Zhejiang	23	29 777	18 643
福 建 Fujian	22	22 892	14 283
山 东 Shandong	47	55 433	37 763
广 东 Guangdong	25	37 014	23 842
广 西 Guangxi	17	16 297	10 919
海 南 Hainan	6	6 597	3 968
其 他 Others	147	177 295	116 904

主要统计指标解释

海洋专业　指高等教育和中等职业教育所设的与海洋有关的专业。

Explanatory Notes on Main Statistical Indicators

Marine Speciality refers to the ocean-related speciality in higher education and the professional secondary vocational education.

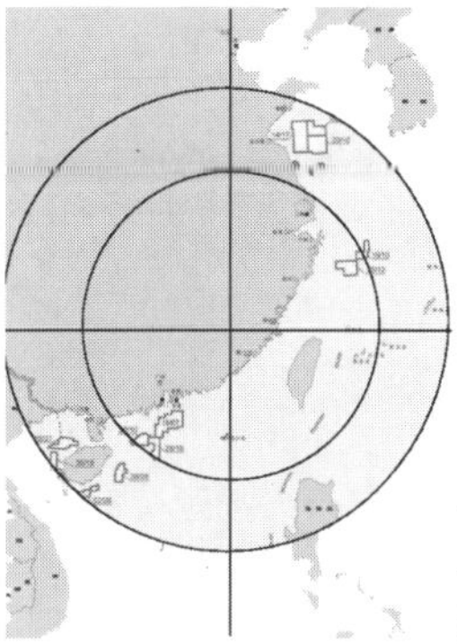

8
海洋环境保护
Marine Environmental Protection

8-1 海区海水水质评价结果
Seawater Quality Assessment Results

项　目 Item	季　节 Season	**合计（平方千米）** **Total（km^2）**	第二类水质海域面积（平方千米） Area of the Second Grade Sea Waters（km^2）	第三类水质海域面积（平方千米） Area of the Third Grade Sea Waters（km^2）
全海域 Total Area	冬季	**211 910**	72 440	40 030
	春季	**163 620**	54 080	34 590
	夏季	**154 610**	54 120	36 900
	秋季	**195 730**	66 970	36 160
渤　海 Bohai Sea	冬季	**47 090**	23 160	10 300
	春季	**31 220**	12 910	8 540
	夏季	**28 910**	12 010	8 090
	秋季	**41 540**	24 810	5 490
黄　海 Yellow Sea	冬季	**42 190**	23 600	7 750
	春季	**36 520**	13 900	8 490
	夏季	**37 760**	15 570	9 490
	秋季	**43 790**	19 750	6 450
东　海 East China Sea	冬季	**102 740**	19 180	13 290
	春季	**77 140**	21 400	11 430
	夏季	**67 130**	22 050	9 410
	秋季	**84 210**	16 080	14 480
南　海 South China Sea	冬季	**19 890**	6 500	8 690
	春季	**18 740**	5 870	6 130
	夏季	**20 810**	4 490	9 910
	秋季	**26 190**	6 330	9 740

8-1 续表 continued

项 目 Item	第四类水质海域面积（平方千米） Area of the Fourth Grade Sea Waters（km^2）	劣于第四类水质海域面积（平方千米） Sea Area of the Sea Waters Inferior to the Fourth Grade（km^2）	首要超标污染物 Prime Pollutants Exceeding the Set Standard
全海域 Total Area	32 290 23 210 23 570 29 370	67 150 51 740 40 020 63 230	无机氮、活性磷酸盐、石油类 Inorganic Nitrogen, Active Phosphate, Oils
渤 海 Bohai Sea	6 430 5 090 4 750 3 910	7 200 4 680 4 060 7 330	无机氮、石油类 Inorganic Nitrogen, Oils
黄 海 Yellow Sea	4 730 5 940 8 020 8 930	6 110 8 190 4 680 8 660	无机氮、活性磷酸盐、石油类 Inorganic Nitrogen, Active Phosphate, Oils
东 海 East China Sea	19 750 10 330 9 000 12 880	50 520 33 980 26 670 40 770	无机氮、活性磷酸盐 Inorganic Nitrogen, Active Phosphate
南 海 South China Sea	1 380 1 850 1 800 3 650	3 320 4 890 4 610 6 470	无机氮、活性磷酸盐、石油类 Inorganic Nitrogen, Active Phosphate, Oils

8-2 海区废弃物海洋倾倒情况
Ocean Dumping of Wastes by Sea Area

海 区 Sea Area	疏浚物 （万立方米） Dredged Materials （10 000 m³）	惰性无机地质废料 （立方米） Inert, Inorganic Geologic Wastes （m³）
合 计 Total	**13 615.0**	**853.3**
渤黄海 Bohai Sea and Yellow Sea	1 702.0	
东 海 East China Sea	9 162.6	
南 海 South China Sea	2 750.4	853.3

8-3 海区海洋石油勘探开发污染物排放入海情况

Discharge of Pollutants into the Sea from Offshore Oil Exploration and Exploitation

海　区 Sea Area	生产污水 （万立方米） Production Sewage （10 000 m^3）	泥浆 （立方米） Sludge （m^3）	钻屑 （立方米） Debris from Drilling （m^3）	机舱污水 （立方米） Sewage from Engineroom （m^3）	食品废弃物 （立方米） Food Wastes （m^3）	生活污水 （立方米） Domestic Sewage （m^3）
合　计 Total	**17 837.7**	**21 541.0**	**44 909.0**	**4 341.0**	**716.0**	**532 583.0**
渤黄海 Bohai Sea and Yellow Sea	616.2	4 101.0	22 632.0			175 257.0
东　海 East China Sea	61.5	540.0	8 206.0			56 000.0
南　海 South China Sea	17 160.0	16 900.0	14 071.0	4 341.0	716.0	301 326.0

8-4 沿海地区工业废水排放及处理情况
Discharge and Treatment of Industrial Waste Water by Coastal Regions

单位:万吨 (10 000 t)

地区 Region		工业废水排放总量 Total Volume of Industrial Waste Water Discharged	直排入海 Discharged Directly into the Sea	工业废水处理量 Industrial Waste Water Treated
总计 Total		**1 105 711.6**	**95 750.6**	**2 524 566.9**
环渤海经济区 Round-the-Bohai Sea Economic Zone	**合计 Total**	**382 663.5**	**30 787.9**	**1 306 764.0**
	辽宁 Liaoning	83 140.3	21 529.3	340 622.8
	河北 Hebei	94 110.4	1 315.7	598 193.2
	天津 Tianjin	18 972.5	0.0	34 268.4
	山东 Shandong	186 440.3	7 942.9	333 679.5
长江三角洲经济区 Yangtze River Delta Economic Zone	**合计 Total**	**400 719.5**	**19 085.4**	**706 024.7**
	江苏 Jiangsu	206 427.5	1 393.0	418 384.0
	上海 Shanghai	46 938.8	9 824.8	61 219.9
	浙江 Zhejiang	147 353.3	7 867.6	226 420.9
海峡西岸经济区 Economic Zone on the West Side of the Taiwan Straits	**合计 Total**	**90 741.4**	**36 330.7**	**143 089.4**
	福建 Fujian	90 741.4	36 330.7	143 089.4
珠江三角洲经济区 Zhujiang River Delta Economic Zone	**合计 Total**	**161 454.8**	**3 119.9**	**188 722.0**
	广东 Guangdong	161 454.8	3 119.9	188 722.0
环北部湾经济区 Round-the-Beibu Gulf Economic Zone	**合计 Total**	**70 132.3**	**6 426.7**	**179 966.8**
	广西 Guangxi	63 253.4	3 175.6	173 492.8
	海南 Hainan	6 879.0	3 251.1	6 474.0

8-5 沿海城市工业废水排放及处理情况
Discharge and Treatment of Industrial Waste Water by Coastal Cities

单位:万吨 (10 000 t)

沿海城市 Coastal City	工业废水排放总量 Total Volume of Industrial Waste Water Discharged	直排入海 Discharged Directly into the Sea	工业废水处理量 Industrial Waste Water Treated
天　津 Tianjin	18 972.5	0.0	34 268.4
唐　山 Tangshan	11 914.0	1 273.3	240 339.0
秦皇岛 Qinhuangdao	7 263.7	42.4	12 648.6
沧　州 Cangzhou	8 926.2	0.0	9 685.6
大　连 Dalian	34 564.7	19 906.9	23 577.4
丹　东 Dandong	3 624.6	0.9	3 998.0
锦　州 Jinzhou	3 851.4	31.0	4 248.2
营　口 Yingkou	3 040.8	1 005.7	3 483.3
盘　锦 Panjin	4 307.5	0.0	4 135.9
葫芦岛 Huludao	2 405.5	584.8	3 487.0
上　海 Shanghai	46 938.8	9 824.8	61 219.9
南　通 Nantong	15 744.9	0.0	14 601.7
连云港 Lianyungang	7 338.9	295.8	29 956.1
盐　城 Yancheng	16 193.2	1 097.3	22 145.9

8-5 续表1 continued

沿海城市 Coastal City	工业废水排放总量 Total Volume of Industrial Waste Water Discharged	直排入海 Discharged Directly into the Sea	工业废水处理量 Industrial Waste Water Treated
杭　州 Hangzhou	33 807.1	0.0	93 693.3
宁　波 Ningbo	16 097.7	6 698.9	30 532.8
温　州 Wenzhou	6 277.8	0.4	5 675.2
嘉　兴 Jiaxing	21 947.2	146.3	26 565.2
绍　兴 Shaoxing	26 068.6	0.0	28 189.2
舟　山 Zhoushan	2 202.2	876.1	1 318.1
台　州 Taizhou	6 251.5	145.9	5 591.0
福　州 Fuzhou	4 439.0	438.6	10 881.9
厦　门 Xiamen	21 398.4	17 106.6	22 225.0
莆　田 Putian	2 644.3	1 100.1	2 525.6
泉　州 Quanzhou	19 184.7	2 296.8	23 934.9
漳　州 Zhangzhou	21 197.5	15 187.8	19 257.2
宁　德 Ningde	1 529.7	77.5	1 743.5
青　岛 Qingdao	10 566.0	2 064.8	43 835.9
东　营 Dongying	7 819.1	0.0	27 235.9
烟　台 Yantai	9 762.1	2 387.2	15 886.3
潍　坊 Weifang	27 402.2	0.0	43 241.8
威　海 Weihai	2 597.8	83.4	3 192.4
日　照 Rizhao	6 735.2	3 407.5	7 158.9
滨　州 Binzhou	22 521.8	0.0	19 514.2

8-5 续表2 continued

沿海城市 Coastal City	工业废水排放总量 Total Volume of Industrial Waste Water Discharged	直排入海 Discharged Directly into the Sea	工业废水处理量 Industrial Waste Water Treated
广 州 Guangzhou	18 958.6	0.0	20 343.9
深 圳 Shenzhen	19 076.8	477.8	12 848.7
珠 海 Zhuhai	5 933.6	407.4	5 123.2
汕 头 Shantou	5 839.6	46.2	5 465.3
江 门 Jiangmen	9 300.3	44.6	12 731.2
湛 江 Zhanjiang	6 049.0	389.7	6 227.1
茂 名 Maoming	4 110.5	592.0	5 937.8
惠 州 Huizhou	8 594.6	723.6	7 969.1
汕 尾 Shanwei	2 073.6	5.4	715.2
阳 江 Yangjiang	2 966.0	433.4	1 854.8
东 莞 Dongguan	20 428.9	0.0	32 350.6
中 山 Zhongshan	7 455.2	0.0	9 405.3
潮 州 Chaozhou	2 544.2	0.0	1 288.2
揭 阳 Jieyang	5 594.3	0.0	4 616.4
北 海 Beihai	1 816.9	849.5	1 811.9
防城港 Fangchenggang	1 432.6	254.8	4 272.3
钦 州 Qinzhou	3 712.1	2 068.2	3 610.9
海 口 Haikou	696.9	23.2	629.1
三 亚 Sanya	52.8	0.0	40.2

8-6 沿海地带工业废水排放及处理情况
Discharge and Treatment of Industrial Waste Water by Coastal Counties

单位:万吨 (10 000 t)

地区 Region	工业废水排放总量 Total Volume of Industrial Waste Water Discharged	直排入海 Discharged Directly into the Sea	工业废水处理量 Industrial Waste Water Treated
天　津　Tianjin	**8 944.6**	**0.0**	**9 624.7**
河　北　Hebei	**7 125.2**	**1.3**	**32 509.2**
唐　山　Tangshan	1 785.8	0.5	22 457.1
秦皇岛　Qinhuangdao	4 434.1	0.8	9 255.6
沧　州　Cangzhou	905.3	0.0	796.4
辽　宁　Liaoning	**26 597.8**	**14 652.2**	**25 477.6**
大　连　Dalian	19 673.0	13 064.3	15 855.4
丹　东　Dandong	564.1	0.9	810.8
锦　州　Jinzhou	1 129.6	0.0	2 122.4
营　口　Yingkou	2 224.1	1 005.7	2 829.1
盘　锦　Panjin	753.1	0.0	660.3
葫芦岛　Huludao	2 254.0	581.2	3 199.5
上　海　Shanghai	**31 840.7**	**9 824.8**	**53 135.0**
江　苏　Jiangsu	**22 967.3**	**1 393.0**	**51 743.2**
南　通　Nantong	7 432.9	0.0	6 701.4
连云港　Lianyungang	5 514.9	295.8	28 417.9
盐　城　Yancheng	10 019.6	1 097.3	16 623.9

8-6 续表1 continued

地区 Region	工业废水排放总量 Total Volume of Industrial Waste Water Discharged	直排入海 Discharged Directly into the Sea	工业废水处理量 Industrial Waste Water Treated
浙　江 Zhejiang	**59 927.4**	**7 591.2**	**65 157.3**
杭　州 Hangzhou	7 481.3	0.0	8 217.6
宁　波 Ningbo	14 841.8	6 568.6	20 207.0
温　州 Wenzhou	4 770.0	0.4	4 308.6
嘉　兴 Jiaxing	7 605.1	0.2	9 192.9
绍　兴 Shaoxing	18 498.0	0.0	17 642.9
舟　山 Zhoushan	2 202.2	876.1	1 318.1
台　州 Taizhou	4 529.1	145.9	4 270.4
福　建 Fujian	**62 222.2**	**35 292.3**	**60 856.5**
福　州 Fuzhou	3 529.0	438.6	9 496.2
厦　门 Xiamen	21 398.4	17 106.6	22 225.0
莆　田 Putian	1 720.4	184.9	1 602.9
泉　州 Quanzhou	16 720.1	2 296.8	21 448.9
漳　州 Zhangzhou	17 829.5	15 187.8	5 022.7
宁　德 Ningde	1 024.8	77.5	1 060.8
山　东 Shandong	**43 319.9**	**4 859.4**	**115 732.1**
青　岛 Qingdao	8 569.6	2 064.8	42 043.7
东　营 Dongying	6 933.3	0.0	26 434.0
烟　台 Yantai	7 647.0	2 387.2	13 903.3
潍　坊 Weifang	11 505.3	0.0	25 512.3
威　海 Weihai	1 870.4	81.6	2 371.4
日　照 Rizhao	853.1	325.9	1 567.2
滨　州 Binzhou	5 941.3	0.0	3 900.2

8-6 续表2 continued

地区 Region	工业废水排放总量 Total Volume of Industrial Waste Water Discharged	直排入海 Discharged Directly into the Sea	工业废水处理量 Industrial Waste Water Treated
广 东 Guangdong	**79 597.2**	**1 076.0**	**93 285.0**
广 州 Guangzhou	10 292.0	0.0	11 187.0
深 圳 Shenzhen	11 923.7	456.3	8 766.4
珠 海 Zhuhai	2 413.2	0.0	2 167.0
汕 头 Shantou	5 807.5	46.2	5 459.1
江 门 Jiangmen	6 420.1	44.6	9 754.1
湛 江 Zhanjiang	5 634.8	69.7	5 702.9
茂 名 Maoming	2 021.9	4.0	3 689.4
惠 州 Huizhou	1 674.1	16.5	1 589.7
汕 尾 Shanwei	1 915.0	5.4	659.9
阳 江 Yangjiang	1 965.0	433.4	1 159.9
东 莞 Dongguan	20 428.9	0.0	32 350.6
中 山 Zhongshan	7 455.2	0.0	9 405.3
潮 州 Chaozhou	1 097.9	0.0	596.2
揭 阳 Jieyang	547.9	0.0	797.6
广 西 Guangxi	**5 605.2**	**3 172.5**	**7 330.2**
北 海 Beihai	1 816.9	849.5	1 811.9
防城港 Fangchenggang	1 024.5	254.8	2 711.9
钦 州 Qinzhou	2 763.8	2 068.2	2 806.5
海 南 Hainan	**717.9**	**216.5**	**643.1**
海 口 Haikou	665.1	0.0	603.0
三 亚 Sanya	52.8	216.5	40.2

注：1.表中数据为沿海地带合计数(表8-9、表8-13同)。

2.沿海地带是指有海岸线的县、县级市、区(包括直辖市和地级市的区)。

Note: 1. The data in the table are the total numbers for the coastal regions (The same for Tables 8-9, 8-13).

2. Coastal Zone refers to the counties, county-level cities and districts with coastlines (including the districts under the municipalities directly under the Central Government and the prefecture-level districts).

8-7 沿海地区一般工业固体废物倾倒丢弃、处理及综合利用情况

Discharge, Treatment and Multipurpose Utilization of Common Industrial Solid Wastes by Coastal Regions

单位:万吨 (10 000 t)

地区 Region		一般工业固体废物倾倒丢弃量 Volume of Common Industrial Solid Wastes Discharged	一般工业固体废物处置量 Volume of Common Industrial Solid Wastes Treated	一般工业固体废物综合利用量 Volume of Common Industrial Solid Wastes Comprehensively Utilized
总 计 Total		**9.1**	**26 424.0**	**79 570.5**
环渤海经济区 Round-the-Bohai Sea Economic Zone	**合 计 Total**	**7.5**	**23 555.1**	**49 761.4**
	辽 宁 Liaoning	7.5	8 067.3	10 028.9
	河 北 Hebei	0.0	14 729.1	19 900.0
	天 津 Tianjin	0.0	21.5	1 524.0
	山 东 Shandong	0.0	737.2	18 308.6
长江三角洲经济区 Yangtze River Delta Economic Zone	**合 计 Total**	**0.0**	**684.9**	**16 266.3**
	江 苏 Jiangsu	0.0	407.4	10 207.0
	上 海 Shanghai	0.0	72.2	1 796.2
	浙 江 Zhejiang	0.0	205.3	4 263.2
海峡西岸经济区 Economic Zone on the West Side of the Taiwan Straits	**合 计 Total**	**0.0**	**1 157.4**	**3 784.3**
	福 建 Fujian	0.0	1 157.4	3 784.3
珠江三角洲经济区 Zhujiang River Delta Economic Zone	**合 计 Total**	**1.1**	**438.8**	**5 102.7**
	广 东 Guangdong	1.1	438.8	5 102.7
环北部湾经济区 Round-the-Beibu Gulf Economic Zone	**合 计 Total**	**0.5**	**587.8**	**4 655.8**
	广 西 Guangxi	0.4	546.2	4 387.7
	海 南 Hainan	0.1	41.6	268.1

8-8 沿海城市一般工业固体废物倾倒丢弃、处理及综合利用情况
Discharge, Treatment and Multipurposed Utilization of Common Industrial Solid Wastes by Coastal Cities

单位:万吨 (10 000 t)

沿海城市 Coastal City	一般工业固体废物倾倒丢弃量 Volume of Common Industrial Solid Wastes Discharged	一般工业固体废物处置量 Volume of Common Industrial Solid Wastes Treated	一般工业固体废物综合利用量 Volume of Common Industrial Solid Wastes Comprehensively Utilized
天　津 Tianjin	0.0	21.5	1 524.0
唐　山 Tangshan	0.0	2 653.7	7 231.0
秦皇岛 Qinhuangdao	0.0	195.5	789.0
沧　州 Cangzhou	0.0	0.3	464.9
大　连 Dalian	0.0	99.8	397.2
丹　东 Dandong	0.0	303.3	251.5
锦　州 Jinzhou	0.0	27.4	274.8
营　口 Yingkou	0.0	1.9	707.3
盘　锦 Panjin	0.0	8.3	187.2
葫芦岛 Huludao	0.0	15.0	451.4
上　海 Shanghai	0.0	72.2	1 796.2
南　通 Nantong	0.0	5.4	518.1
连云港 Lianyungang	0.0	11.5	583.3
盐　城 Yancheng	0.0	24.7	535.6

8-8　续表1 continued

沿海城市 Coastal City	一般工业固体废物倾倒丢弃量 Volume of Common Industrial Solid Wastes Discharged	一般工业固体废物处置量 Volume of Common Industrial Solid Wastes Treated	一般工业固体废物综合利用量 Volume of Common Industrial Solid Wastes Comprehensively Utilized
杭　州　Hangzhou	0.0	73.5	575.2
宁　波　Ningbo	0.0	28.6	1 117.9
温　州　Wenzhou	0.0	4.8	269.5
嘉　兴　Jiaxing	0.0	30.0	501.2
绍　兴　Shaoxing	0.0	9.8	357.3
舟　山　Zhoushan	0.0	0.5	132.6
台　州　Taizhou	0.0	4.8	243.1
福　州　Fuzhou	0.0	27.8	573.7
厦　门　Xiamen	0.0	13.4	89.6
莆　田　Putian	0.0	8.4	59.7
泉　州　Quanzhou	0.0	7.3	795.2
漳　州　Zhangzhou	0.0	6.9	259.9
宁　德　Ningde	0.0	41.1	412.9
青　岛　Qingdao	0.0	28.9	664.5
东　营　Dongying	0.0	11.7	342.7
烟　台　Yantai	0.0	305.5	1 905.0
潍　坊　Weifang	0.0	18.3	1 081.2
威　海　Weihai	0.0	18.2	330.2
日　照　Rizhao	0.0	6.1	810.5
滨　州　Binzhou	0.0	21.9	2 090.4

8-8 续表2 continued

沿海城市 Coastal City	一般工业固体废物倾倒丢弃量 Volume of Common Industrial Solid Wastes Discharged	一般工业固体废物处置量 Volume of Common Industrial Solid Wastes Treated	一般工业固体废物综合利用量 Volume of Common Industrial Solid Wastes Comprehensively Utilized
广 州 Guangzhou	0.0	20.8	436.0
深 圳 Shenzhen	0.0	40.5	52.4
珠 海 Zhuhai	0.0	16.4	276.0
汕 头 Shantou	0.0	2.2	105.2
江 门 Jiangmen	0.0	33.4	190.6
湛 江 Zhanjiang	0.0	1.6	221.0
茂 名 Maoming	0.0	4.8	204.0
惠 州 Huizhou	0.0	3.2	70.9
汕 尾 Shanwei	0.0	0.4	111.1
阳 江 Yangjiang	0.7	0.1	473.1
东 莞 Dongguan	0.1	55.2	433.2
中 山 Zhongshan	0.0	6.8	86.4
潮 州 Chaozhou	0.0	0.1	104.0
揭 阳 Jieyang	0.1	1.0	114.5
北 海 Beihai	0.1	3.8	296.9
防城港 Fangchenggang	0.0	0.1	259.8
钦 州 Qinzhou	0.0	1.6	99.4
海 口 Haikou	0.0	0.5	4.1
三 亚 Sanya	0.0	8.3	0.0

8-9 沿海地带一般工业固体废物倾倒丢弃、处理及综合利用情况
Discharge, Treatment and Multipurposed Utilization of Common Industrial Solid Wastes by Coastal Counties

单位:万吨　　　　(10 000 t)

地区 Region	一般工业固体废物倾倒丢弃量 Volume of Common Industrial Solid Wastes Discharged	一般工业固体废物处置量 Volume of Common Industrial Solid Wastes Treated	一般工业固体废物综合利用量 Volume of Common Industrial Solid Wastes Comprehensively Utilized
天　津　Tianjin	**0.0**	**17.3**	**418.5**
河　北　Hebei	**0.0**	**195.7**	**2 205.9**
唐　山　Tangshan	0.0	5.3	1 514.7
秦皇岛　Qinhuangdao	0.0	190.4	689.1
沧　州　Cangzhou	0.0	0.0	2.2
辽　宁　Liaoning	**0.0**	**99.1**	**1 283.5**
大　连　Dalian	0.0	80.9	324.5
丹　东　Dandong	0.0	0.0	41.9
锦　州　Jinzhou	0.0	2.5	17.6
营　口　Yingkou	0.0	1.9	672.1
盘　锦　Panjin	0.0	0.0	13.2
葫芦岛　Huludao	0.0	13.8	214.1
上　海　Shanghai	**0.0**	**45.3**	**1 635.0**
江　苏　Jiangsu	**0.0**	**27.8**	**1 171.4**
南　通　Nantong	0.0	1.9	220.3
连云港　Lianyungang	0.0	10.2	487.3
盐　城　Yancheng	0.0	15.7	463.8

8-9 续表1 continued

地区 Region	一般工业固体废物倾倒丢弃量 Volume of Common Industrial Solid Wastes Discharged	一般工业固体废物处置量 Volume of Common Industrial Solid Wastes Treated	一般工业固体废物综合利用量 Volume of Common Industrial Solid Wastes Comprehensively Utilized
浙 江 Zhejiang	**0.0**	**55.9**	**2 259.2**
杭 州 Hangzhou	0.0	12.7	98.1
宁 波 Ningbo	0.0	27.0	1 083.5
温 州 Wenzhou	0.0	3.6	232.6
嘉 兴 Jiaxing	0.0	1.8	217.8
绍 兴 Shaoxing	0.0	5.7	277.7
舟 山 Zhoushan	0.0	0.5	132.6
台 州 Taizhou	0.0	4.7	216.9
福 建 Fujian	**0.0**	**52.7**	**1 438.2**
福 州 Fuzhou	0.0	26.9	467.6
厦 门 Xiamen	0.0	13.4	89.6
莆 田 Putian	0.0	0.1	33.5
泉 州 Quanzhou	0.0	6.3	354.6
漳 州 Zhangzhou	0.0	3.2	163.0
宁 德 Ningde	0.0	2.9	329.9
山 东 Shandong	**0.0**	**371.9**	**3 945.4**
青 岛 Qingdao	0.0	28.2	455.9
东 营 Dongying	0.0	11.7	267.9
烟 台 Yantai	0.0	304.5	1 829.5
潍 坊 Weifang	0.0	1.8	496.8
威 海 Weihai	0.0	17.7	138.7
日 照 Rizhao	0.0	1.1	489.5
滨 州 Binzhou	0.0	6.8	267.1

8-9 续表2 continued

地区 Region	一般工业固体废物倾倒丢弃量 Volume of Common Industrial Solid Wastes Discharged	一般工业固体废物处置量 Volume of Common Industrial Solid Wastes Treated	一般工业固体废物综合利用量 Volume of Common Industrial Solid Wastes Comprehensively Utilized
广 东 Guangdong	**0.5**	**163.1**	**2 099.1**
广 州 Guangzhou	0.0	14.9	355.7
深 圳 Shenzhen	0.0	39.0	50.8
珠 海 Zhuhai	0.0	8.6	3.8
汕 头 Shantou	0.0	2.2	105.2
江 门 Jiangmen	0.0	27.5	168.3
湛 江 Zhanjiang	0.0	1.1	175.3
茂 名 Maoming	0.0	4.8	107.2
惠 州 Huizhou	0.0	2.1	30.1
汕 尾 Shanwei	0.0	0.4	63.5
阳 江 Yangjiang	0.4	0.0	333.8
东 莞 Dongguan	0.1	55.2	433.2
中 山 Zhongshan	0.0	6.8	86.4
潮 州 Chaozhou	0.0	0.1	100.5
揭 阳 Jieyang	0.0	0.3	85.1
广 西 Guangxi	**0.1**	**4.5**	**552.4**
北 海 Beihai	0.1	3.8	296.9
防城港 Fangchenggang	0.0	0.1	195.7
钦 州 Qinzhou	0.0	0.6	59.9
海 南 Hainan	**0.0**	**8.8**	**3.7**
海 口 Haikou	0.0	0.5	3.7
三 亚 Sanya	0.0	8.3	0.0

8-10 主要沿海城市工业废气排放及处理情况
Emission and Treatment of Industrial Waste Gas in Major Coastal Cities

城市 City	工业废气排放量（亿立方米）Industrial Waste Gas Emission（100 000 000 m³）	工业二氧化硫排放量（吨）Industrial Sulphur Dioxide Emission（t）	工业二氧化硫去除量（吨）Industrial Sulphur Dioxide Treated（t）	工业氮氧化物排放量（吨）Industrial Nitrogen Oxide Emission（t）	工业氮氧化物处理量（吨）Industrial Nitrogen Oxide Treated（t）
天津 Tianjin	8 355.2	154 605.0	367 668.4	150 209.6	93 366.9
唐山 Tangshan	35 891.5	214 723.2	413 294.5	228 451.8	81 021.8
秦皇岛 Qinhuangdao	3 391.6	46 688.5	74 193.1	41 223.1	22 457.7
沧州 Cangzhou	2 864.0	32 712.5	118 816.2	27 710.0	23 409.9
大连 Dalian	2 846.8	95 796.1	198 340.8	83 523.9	16 332.2
丹东 Dandong	2 322.5	32 720.8	49 528.2	16 209.0	16 109.3
锦州 Jinzhou	928.4	42 437.5	42 003.5	14 522.8	8 408.1
营口 Yingkou	4 985.0	46 052.4	58 135.0	41 986.6	18 882.5
盘锦 Panjin	977.4	50 123.4	35 265.1	22 840.3	8 004.3
葫芦岛 Huludao	1 374.7	53 752.5	640 071.3	37 452.9	27 371.3
上海 Shanghai	12 801.7	104 851.8	409 097.0	121 492.2	112 340.3
南通 Nantong	2 451.7	55 061.8	205 975.2	43 586.6	21 456.2
连云港 Lianyungang	3 117.9	41 579.2	54 742.2	28 781.9	10 255.0
盐城 Yancheng	2 701.6	41 337.9	141 662.9	26 161.3	27 731.4
杭州 Hangzhou	4 204.4	63 814.5	76 285.5	55 972.9	19 679.3
宁波 Ningbo	6 269.0	101 979.6	405 393.3	131 615.9	78 951.7
温州 Wenzhou	4 226.7	37 316.4	94 413.0	31 735.5	15 323.0

8-10 续表1 continued

城市 City	工业废气排放量（亿立方米） Industrial Waste Gas Emission（100 000 000 m^3）	工业二氧化硫排放量（吨） Industrial Sulphur Dioxide Emission（t）	工业二氧化硫去除量（吨） Industrial Sulphur Dioxide Treated（t）	工业氮氧化物排放量（吨） Industrial Nitrogen Oxide Emission（t）	工业氮氧化物处理量（吨） Industrial Nitrogen Oxide Treated（t）
嘉 兴 Jiaxing	2 240.6	67 924.0	110 201.1	40 060.6	26 783.2
绍 兴 Shaoxing	1 645.4	59 979.7	69 473.1	39 789.5	9 500.1
舟 山 Zhoushan	539.2	12 379.5	56 480.8	11 554.4	8 081.6
台 州 Taizhou	1 319.0	31 867.6	112 086.5	26 254.7	5 369.3
福 州 Fuzhou	3 134.8	55 370.4	120 775.3	64 751.0	35 699.3
厦 门 Xiamen	953.4	17 028.1	30 676.6	11 778.0	13 031.6
莆 田 Putian	503.1	11 081.6	21 898.3	11 214.2	5 165.5
泉 州 Quanzhou	3 794.2	94 698.6	206 112.9	69 431.9	19 515.4
漳 州 Zhangzhou	1 772.5	35 538.0	67 903.4	38 373.5	32 075.4
宁 德 Ningde	1 379.9	17 678.6	35 203.4	15 453.3	27 122.3
青 岛 Qingdao	2 213.7	65 003.4	231 029.0	63 680.1	27 189.6
东 营 Dongying	1 346.0	47 859.5	408 546.9	27 669.2	15 100.1
烟 台 Yantai	4 052.0	67 531.2	490 820.7	57 240.6	43 607.8
潍 坊 Weifang	4 454.9	106 598.6	348 119.2	59 941.7	31 993.5
威 海 Weihai	913.3	24 662.6	102 708.8	24 210.8	23 338.8
日 照 Rizhao	5 070.6	42 127.8	138 302.4	42 705.5	30 920.5
滨 州 Binzhou	5 901.1	93 908.9	1 434 969.3	97 082.5	115 252.5
广 州 Guangzhou	3 550.9	48 840.8	423 826.1	44 349.1	49 396.9

8-10 续表2 continued

城 市 City	工业废气排放量（亿立方米） Industrial Waste Gas Emission (100 000 000 m³)	工业二氧化硫排放量（吨） Industrial Sulphur Dioxide Emission (t)	工业二氧化硫去除量（吨） Industrial Sulphur Dioxide Treated (t)	工业氮氧化物排放量（吨） Industrial Nitrogen Oxide Emission (t)	工业氮氧化物处理量（吨） Industrial Nitrogen Oxide Treated (t)
深 圳 Shenzhen	2 312.2	4 131.9	32 115.3	10 672.9	14 921.0
珠 海 Zhuhai	2 045.6	21 946.4	65 938.0	26 000.0	20 106.8
汕 头 Shantou	798.9	22 423.4	49 374.3	14 911.1	8 898.3
江 门 Jiangmen	1 372.4	43 703.5	95 605.1	42 157.7	26 156.6
湛 江 Zhanjiang	1 078.3	29 599.0	74 593.5	15 813.3	9 105.3
茂 名 Maoming	915.4	31 792.8	438 104.0	16 712.0	4 919.7
惠 州 Huizhou	1 611.5	28 683.1	33 494.4	34 682.4	22 124.9
汕 尾 Shanwei	647.3	12 807.3	76 717.1	5 631.5	11 571.6
阳 江 Yangjiang	1 217.2	24 066.2	56 361.6	21 989.8	10 696.2
东 莞 Dongguan	3 218.8	103 309.9	185 613.8	71 824.8	81 036.3
中 山 Zhongshan	852.3	24 429.8	8 744.4	18 371.0	1 459.5
潮 州 Chaozhou	710.9	12 867.8	54 873.5	8 757.6	11 708.7
揭 阳 Jieyang	702.7	23 385.5	72 405.9	8 659.1	10 895.1
北 海 Beihai	748.6	12 184.0	24 766.6	6 364.4	1 970.7
防城港 Fangcheng-gang	576.6	27 854.7	37 436.7	14 788.8	2 290.2
钦 州 Qinzhou	518.9	15 967.1	23 604.6	7 082.6	5 725.8
海 口 Haikou	78.0	2 517.4	986.6	198.9	1.1
三 亚 Sanya	48.4	207.7	821.4	675.6	742.9

8-10 续表3 continued

城 市 City	工业烟(粉)尘排放量(亿立方米) Industrial Soot (Dust) Emission (100 000 000 m^3)	工业烟粉尘处理量(吨) Industrial Soot (Dust) Treated (t)	生活二氧化硫排放量(吨) Household Sulphur Dioxide Emission (t)	生活氮氧化物排放量(吨) Household Nitrogen Oxide Emission (t)	生活烟尘排放量(吨) Household Soot Emission (t)
天 津 Tianjin	20 975.0	1 883 132.8	385.0	93.0	45.0
唐 山 Tangshan	32 588.3	1 462 965.9	734.4	201.6	270.0
秦皇岛 Qinhuangdao	3 049.6	566 733.3	288.0	98.3	180.6
沧 州 Cangzhou	16 263.1	1 701 005.0	788.6	136.4	93.7
大 连 Dalian	90 911.3	2 259 171.9	1 279.0	169.0	547.0
丹 东 Dandong	2 414.0	379 750.2	180.9	65.6	85.1
锦 州 Jinzhou	4 908.8	204 834.5	1 406.2	451.7	1 054.7
营 口 Yingkou	84 653.3	1 996 240.6	3 195.2	557.7	1 769.0
盘 锦 Panjin	20 687.1	712 705.5	795.3	117.0	467.8
葫芦岛 Huludao	8 472.5	516 989.1	846.3	70.0	534.4
上 海 Shanghai	28 767.1	2 057 343.6	27 088.6	2 647.9	9 805.8
南 通 Nantong	5 155.7	1 474 179.3	1 720.9	864.7	925.5
连云港 Lianyungang	30 315.8	4 502 953.8	9 837.0	2 058.8	6 651.3
盐 城 Yancheng	55 380.2	5 639 869.6	12 732.9	1 650.6	10 206.7
杭 州 Hangzhou	12 541.6	1 308 652.8	11 908.3	1 352.9	2 956.2
宁 波 Ningbo	103 306.0	3 578 992.0	16 633.0	3 255.3	14 900.0
温 州 Wenzhou	30 122.7	10 171 866.9	3 642.6	555.6	2 610.9

8-10 续表4 continued

城 市 City	工业烟(粉)尘排放量(亿立方米) Industrial Soot (Dust) Emission (100 000 000 m^3)	工业烟粉尘处理量(吨) Industrial Soot (Dust) Treated (t)	生活二氧化硫排放量(吨) Household Sulphur Dioxide Emission (t)	生活氮氧化物排放量(吨) Household Nitrogen Oxide Emission (t)	生活烟尘排放量(吨) Household Soot Emission (t)
嘉 兴 Jiaxing	20 975.0	1 883 132.8	385.0	93.0	45.0
绍 兴 Shaoxing	32 588.3	1 462 965.9	734.4	224.0	144.0
舟 山 Zhoushan	3 049.6	566 733.3	288.0	98.3	180.6
台 州 Taizhou	16 263.1	1 701 005.0	1 036.7	160.1	108.9
福 州 Fuzhou	90 911.3	2 259 171.9	1 726.0	306.0	1 096.0
厦 门 Xiamen	2 414.0	379 750.2	265.5	92.0	85.1
莆 田 Putian	4 908.8	204 834.5	1 702.8	530.6	1 370.7
泉 州 Quanzhou	84 653.3	1 996 240.6	3 820.3	752.6	2 115.1
漳 州 Zhangzhou	20 687.1	712 705.5	817.0	96.1	480.6
宁 德 Ningde	8 472.5	516 989.1	1 112.5	70.0	534.4
青 岛 Qingdao	28 767.1	2 057 343.6	27 088.6	2 647.9	9 805.8
东 营 Dongying	5 155.7	1 474 179.3	1 720.9	864.7	925.5
烟 台 Yantai	30 315.8	4 502 953.8	15 664.2	5 369.0	6 457.8
潍 坊 Weifang	55 380.2	5 639 869.6	15 461.0	2 000.8	7 011.7
威 海 Weihai	12 541.6	1 308 652.8	17 190.1	2 701.6	2 956.2
日 照 Rizhao	103 306.0	3 578 992.0	16 721.8	5 704.9	12 900.0
滨 州 Binzhou	30 122.7	10 171 866.9	3 661.2	1 525.5	6 680.7
广 州 Guangzhou	9 297.7	3 354 930.9	2 363.0	1 963.0	214.0

8-10 续表5 continued

城 市 City	工业烟(粉)尘排放量(亿立方米) Industrial Soot (Dust) Emission (100 000 000 m³)	工业烟粉尘处理量(吨) Industrial Soot (Dust) Treated (t)	生活二氧化硫排放量(吨) Household Sulphur Dioxide Emission (t)	生活氮氧化物排放量(吨) Household Nitrogen Oxide Emission (t)	生活烟尘排放量(吨) Household Soot Emission (t)
深 圳 Shenzhen	1 078.7	266 417.0	238.0	885.0	82.0
珠 海 Zhuhai	10 446.3	822 534.6	0.0	0.0	0.0
汕 头 Shantou	4 351.6	598 592.0	134.0	28.6	73.0
江 门 Jiangmen	13 031.3	1397 594.2	110.0	7.0	70.0
湛 江 Zhanjiang	9 361.4	1001 627.7	2 350.0	366.9	239.6
茂 名 Maoming	11 476.0	695 585.1	2 392.8	215.4	501.6
惠 州 Huizhou	15 714.6	2196 564.6	239.0	83.4	67.3
汕 尾 Shanwei	3 656.8	560 896.0	624.3	755.0	480.8
阳 江 Yangjiang	19 577.3	599 927.7	823.0	242.0	207.0
东 莞 Dongguan	14 411.5	1532 805.8	1 560.0	156.0	450.0
中 山 Zhongshan	12 276.0	311 929.4	276.6	52.9	234.7
潮 州 Chaozhou	2 213.6	629 256.6	1 027.2	4.0	3.0
揭 阳 Jieyang	3 424.9	268 591.0	4 318.0	635.0	2 540.0
北 海 Beihai	4 692.0	250 963.0	1 282.0	145.4	408.5
防城港 Fangcheng-gang	48 783.6	1224 796.3	971.0	104.0	406.5
钦 州 Qinzhou	4 624.0	200 695.6	1 117.0	76.7	237.3
海 口 Haikou	854.4	2 602.7	19.5	26.8	240.7
三 亚 Sanya	935.1	62 936.9	71.6	13.7	7.0

8-11 沿海地区污染治理项目情况
Pollution Treatment Projects in Coastal Regions

单位:个 (unit)

地区 Region		当年安排施工项目 Arranged for Construction in the Year		当年竣工项目 Completed in the Year	
		治理废水 Treatment of Waste Water	治理固体废物 Treatment of Solid Wastes	治理废水 Treatment of Waste Water	治理固体废物 Treatment of Solid Wastes
总 计 Total		**1 074**	**124**	**793**	**105**
环渤海经济区 Round-the-Bohai Sea Economic Zone	**合 计 Total**	**151**	**23**	**117**	**17**
	辽 宁 Liaoning	19	11	13	8
	河 北 Hebei	16	1	13	1
	天 津 Tianjin	23	2	18	2
	山 东 Shandong	93	9	73	6
长江三角洲经济区 Yangtze River Delta Economic Zone	**合 计 Total**	**607**	**26**	**463**	**22**
	江 苏 Jiangsu	219	6	173	4
	上 海 Shanghai	79	1	62	1
	浙 江 Zhejiang	309	19	228	17
海峡西岸经济区 Economic Zone on the West Side of the Taiwan Straits	**合 计 Total**	**89**	**62**	**45**	**56**
	福 建 Fujian	89	62	45	56
珠江三角洲经济区 Zhujiang River Delta Economic Zone	**合 计 Total**	**188**	**6**	**138**	**5**
	广 东 Guangdong	188	6	138	5
环北部湾经济区 Round-the-Beibu Gulf Economic Zone	**合 计 Total**	**39**	**7**	**30**	**5**
	广 西 Guangxi	36	7	29	5
	海 南 Hainan	3	0	1	0

8-12 沿海城市污染治理项目情况
Pollution Treatment Projects in Coastal Cities

单位：个 (unit)

沿海城市 Coastal City	当年安排施工项目 Arranged for Construction in the Year		当年竣工项目 Completed in the Year	
	治理废水 Treatment of Waste Water	治理固体废物 Treatment of Solid Wastes	治理废水 Treatment of Waste Water	治理固体废物 Treatment of Solid Wastes
天 津 Tianjin	23	2	18	2
唐 山 Tangshan	0	0	0	0
秦皇岛 Qinhuangdao	0	1	0	1
沧 州 Cangzhou	1	0	1	0
大 连 Dalian	0	0	0	0
丹 东 Dandong	0	0	0	0
锦 州 Jinzhou	4	0	2	0
营 口 Yingkou	0	0	0	0
盘 锦 Panjin	1	6	1	5
葫芦岛 Huludao	3	0	2	0
上 海 Shanghai	79	1	62	1
南 通 Nantong	21	3	19	2
连云港 Lianyungang	29	0	24	0
盐 城 Yancheng	57	1	43	1

8-12 续表1 continued

沿海城市 Coastal City	当年安排施工项目 Arranged for Construction in the Year		当年竣工项目 Completed in the Year	
	治理废水 Treatment of Waste Water	治理固体废物 Treatment of Solid Wastes	治理废水 Treatment of Waste Water	治理固体废物 Treatment of Solid Wastes
杭 州 Hangzhou	12	2	11	2
宁 波 Ningbo	24	1	14	0
温 州 Wenzhou	5	0	5	0
嘉 兴 Jiaxing	56	4	45	4
绍 兴 Shaoxing	51	4	31	3
舟 山 Zhoushan	11	0	8	0
台 州 Taizhou	42	2	29	2
福 州 Fuzhou	2	1	1	1
厦 门 Xiamen	8	1	5	1
莆 田 Putian	0	0	0	0
泉 州 Quanzhou	37	55	7	51
漳 州 Zhangzhou	19	1	14	1
宁 德 Ningde	1	0	1	0
青 岛 Qingdao	0	0	0	0
东 营 Dongying	7	0	3	0
烟 台 Yantai	19	4	15	3
潍 坊 Weifang	5	0	4	0
威 海 Weihai	4	0	4	0
日 照 Rizhao	1	0	1	0
滨 州 Binzhou	6	0	4	0

8-12 续表2 continued

沿海城市 Coastal City	当年安排施工项目 Arranged for Construction in the Year		当年竣工项目 Completed in the Year	
	治理废水 Treatment of Waste Water	治理固体废物 Treatment of Solid Wastes	治理废水 Treatment of Waste Water	治理固体废物 Treatment of Solid Wastes
广 州 Guangzhou	13	0	10	0
深 圳 Shenzhen	5	0	1	0
珠 海 Zhuhai	9	0	6	0
汕 头 Shantou	20	1	19	1
江 门 Jiangmen	28	0	25	0
湛 江 Zhanjiang	29	0	10	0
茂 名 Maoming	3	0	3	0
惠 州 Huizhou	8	0	7	0
汕 尾 Shanwei	1	0	1	0
阳 江 Yangjiang	6	0	6	0
东 莞 Dongguan	14	1	11	1
中 山 Zhongshan	1	0	0	0
潮 州 Chaozhou	0	0	0	0
揭 阳 Jieyang	0	0	0	0
北 海 Beihai	2	0	2	0
防城港 Fangchenggang	1	0	1	0
钦 州 Qinzhou	1	0	1	0
海 口 Haikou	0	0	0	0
三 亚 Sanya	0	0	0	0

8-13 沿海地带污染治理项目情况
Pollution Treatment Projects in Coastal Counties

单位:个 (unit)

地 区 Region	当年安排施工项目 Arranged for Construction in the Year		当年竣工项目 Completed in the Year	
	治理废水 Treatment of Waste Water	治理固体废物 Treatment of Solid Wastes	治理废水 Treatment of Waste Water	治理固体废物 Treatment of Solid Wastes
天 津 Tianjin	**10**	**1**	**8**	**1**
河 北 Hebei	**0**	**1**	**0**	**1**
唐 山 Tangshan	0	0	0	0
秦皇岛 Qinhuangdao	0	1	0	1
沧 州 Cangzhou	0	0	0	0
辽 宁 Liaoning	**3**	**0**	**2**	**0**
大 连 Dalian	0	0	0	0
丹 东 Dandong	0	0	0	0
锦 州 Jinzhou	0	0	0	0
营 口 Yingkou	0	0	0	0
盘 锦 Panjin	0	0	0	0
葫芦岛 Huludao	3	0	2	0
上 海 Shanghai	**59**	**1**	**44**	**1**
江 苏 Jiangsu	**106**	**4**	**85**	**3**
南 通 Nantong	20	3	18	2
连云港 Lianyungang	29	0	24	0
盐 城 Yancheng	57	1	43	1

8-13 续表1 continued

地 区 Region	当年安排施工项目 Arranged for Construction in the Year		当年竣工项目 Completed in the Year	
	治理废水 Treatment of Waste Water	治理固体废物 Treatment of Solid Wastes	治理废水 Treatment of Waste Water	治理固体废物 Treatment of Solid Wastes
浙 江 Zhejiang	**128**	**8**	**91**	**6**
杭 州 Hangzhou	7	0	7	0
宁 波 Ningbo	24	1	14	0
温 州 Wenzhou	5	0	5	0
嘉 兴 Jiaxing	26	4	21	4
绍 兴 Shaoxing	16	2	9	1
舟 山 Zhoushan	11	0	8	0
台 州 Taizhou	39	1	27	1
福 建 Fujian	**55**	**11**	**19**	**8**
福 州 Fuzhou	2	1	1	1
厦 门 Xiamen	8	1	5	1
莆 田 Putian	0	0	0	0
泉 州 Quanzhou	36	8	6	5
漳 州 Zhangzhou	8	1	6	1
宁 德 Ningde	1	0	1	0
山 东 Shandong	**36**	**4**	**27**	**3**
青 岛 Qingdao	0	0	0	0
东 营 Dongying	7	0	3	0
烟 台 Yantai	19	4	15	3
潍 坊 Weifang	2	0	2	0
威 海 Weihai	3	0	3	0
日 照 Rizhao	0	0	0	0
滨 州 Binzhou	5	0	4	0

8-13 续表2 continued

地 区 Region	当年安排施工项目 Arranged for Construction in the Year		当年竣工项目 Completed in the Year	
	治理废水 Treatment of Waste Water	治理固体废物 Treatment of Solid Wastes	治理废水 Treatment of Waste Water	治理固体废物 Treatment of Solid Wastes
广 东 Guangdong	**82**	**1**	**70**	**1**
广 州 Guangzhou	11	0	9	0
深 圳 Shenzhen	1	0	1	0
珠 海 Zhuhai	0	0	0	0
汕 头 Shantou	20	0	19	0
江 门 Jiangmen	14	0	11	0
湛 江 Zhanjiang	12	0	10	0
茂 名 Maoming	1	0	1	0
惠 州 Huizhou	1	0	1	0
汕 尾 Shanwei	1	0	1	0
阳 江 Yangjiang	6	0	6	0
东 莞 Dongguan	14	1	11	1
中 山 Zhongshan	1	0	0	0
潮 州 Chaozhou	0	0	0	0
揭 阳 Jieyang	0	0	0	0
广 西 Guangxi	**3**	**0**	**3**	**0**
北 海 Beihai	2	0	2	0
防城港 Fangchenggang	0	0	0	0
钦 州 Qinzhou	1	0	1	0
海 南 Hainan	**0**	**0**	**0**	**0**
海 口 Haikou	0	0	0	0
三 亚 Sanya	0	0	0	0

8-14 沿海区域海洋类型自然保护区建设情况
Construction of Marine-Type Nature Reserves in Coastal Regions

地 区 Region	保护区数量(个) Number of Nature Reserves	按保护级别分(个) By Level of Protection		按保护类型分(个) By Type of Protection				保护区面积(平方千米) Area (km^2)
		国家级 National	地方级 Provincial	海洋和海岸生态系统 Marine and Coastal Ecosystem	海洋自然历史遗迹 Marine Natural and Historical Relics	海洋生物多样性 Marine Biodiversity	其他 Others	
合 计 Total	**183**	**43**	**140**	**59**	**25**	**59**	**40**	**250 182**
环渤海经济区 Round-the-Bohai Sea Economic Zone	50	15	35	26	3	21	0	16 318
长江三角洲经济区 Yangtze River Delta Economic Zone	13	7	6	3	1	4	5	21 184
海峡西岸经济区 Economic Zone on the West Side of the Taiwan Straits	48	10	38	8	2	5	33	183 753
珠江三角洲经济区 Zhujiang River Delta Economic Zone	50	5	45	20	0	28	2	3 820
环北部湾经济区 Round-the-Beibu Gulf Economic Zone	22	6	16	2	19	1	0	25 107

8-15 沿海地区海洋类型自然保护区建设情况 Construction of Marine-Type Nature Reserves

地区 Region	保护区数量(个) Number of Nature Reserves	按保护级别分(个) By Level of Protection		按保护类型分(个) By Type of Protection				保护区面积(平方千米) Area (km^2)
		国家级 National	地方级 Provincial	海洋和海岸生态系统 Marine and Coastal Ecosystem	海洋自然历史遗迹 Marine Natural and Historical Relics	海洋生物多样性 Marine Biodiversity	其他 Others	
合计 Total	**183**	**43**	**140**	**59**	**25**	**59**	**40**	**250 182**
天津 Tianjin	1	1	0	1	0	0	0	359
河北 Hebei	3	1	2	2	1	0	0	344
辽宁 Liaoning	15	5	10	10	2	3	0	9 860
上海 Shanghai	4	2	2	0	0	1	3	941
江苏 Jiangsu	3	3	0	0	0	2	1	19 517
浙江 Zhejiang	6	2	4	3	1	1	1	726
福建 Fujian	48	10	38	8	2	5	33	183 753
山东 Shandong	31	8	23	13	0	18	0	5 755
广东 Guangdong	50	5	45	20	0	28	2	3 820
广西 Guangxi	2	2	0	2	0	0	0	110
海南 Hainan	20	4	16	0	19	1	0	24 997

8-16 全国海洋生态监控区基本情况
Basic Condition of the Marine Ecological Monitoring Areas Throughout the Country

生态监控区 Ecological Monitoring Area	所在地 Location	面积（平方千米）Area (km^2)	主要生态系统类型 Major Types of Ecosystem	多样性指数 Diversity Indices		
				浮游植物 Phyto-plankton	大型浮游动物 Macrozoo-plankton	底栖生物 Macrobenthos
双台子河口 Shungtaizi Estuary	辽宁省 Liaoning Province	3 000	河口 Estuary	2.7	1.6	
锦州湾 Jinzhou Bay	辽宁省 Liaoning Province	650	海湾 Bay	2.4	1.1	
滦河口-北戴河 Luanhe River Mouth-Beidaihe	河北省 Hebei Province	900	河口 Estuary	2.7	1.5	1.5
渤海湾 Bohai Bay	天津市 Tianjin Municipality	3 000	海湾 Bay	3.0	2.3	2.9
莱州湾 Laizhou Bay	山东省 Shandong Province	3 770	海湾 Bay	2.1	1.5	2.1
黄河口 Yellow River Mouth	山东省 Shandong Province	2 600	河口 Estuary	2.8	1.0	2.3
苏北浅滩 North Jiangsu Bank	江苏省 Jiangsu Province	15 400	滩涂湿地 Mudflat and Wetland	3.4	1.9	1.7
长江口 Yangtze River Mouth	上海市 Shanghai Municipality	13 668	河口 Estuary	2.2	2.4	1.7
杭州湾 Hangzhou Bay	上海市 浙江省 Shanghai Municipality Zhejiang Province	5 000	海湾 Bay	1.8	1.3	0.3
乐清湾 Yueqing Bay	浙江省 Zhejiang Province	464	海湾 Bay	1.9	2.8	1.6

8-16 续表 continued

生态监控区 Ecological Monitoring Area	所在地 Location	面积（平方千米）Area（km^2）	主要生态系统类型 Major Types of Ecosystem	多样性指数 Diversity Indices		
				浮游植物 Phyto-plankton	大型浮游动物 Macrozoo-plankton	底栖生物 Macrobenthos
闽东沿岸 Coastal East Fujian	福建省 Fujian Province	5 063	海湾 Bay	1.6	2.9	2.5
大亚湾 Daya Bay	广东省 Guangdong Province	1 200	海湾 Bay	2.0	2.9	1.9
珠江口 Zhujiang River Mouth	广东省 Guangdong Province	3 980	河口 Estuary	2.2	4.0	1.4
雷州半岛西南沿岸 Southwest Coast of Leizhou Peninsula	广东省 Guangdong Province	1 150	珊瑚礁 Coral Reef			
广西北海 Beihai, Guangxi	广西壮族自治区 Guangxi Zhuang Nationality Autonomous Region	120	珊瑚礁 红树林 海草床 Coral Reef Mangroves Seagrass Bed			
北仑河口 Beilun River Mouth	广西壮族自治区 Guangxi Zhuang Nationality Autonomous Region	150	红树林 Mangroves			
海南东海岸 East Coast of Hainan	海南省 Hainan Province	3 750	珊瑚礁 海草床 Coral Reef Seagrass Bed			
西沙珊瑚礁 Xisha Coral Reef	海南省 Hainan Province	400	珊瑚礁 Coral Reef			

注：生物多样性指数是生物种数和种类间个体数量分配均匀性的综合表现，用Shannon-Wiener多样性指数表征。

Note: Biodiversity index refers to the comprehensive expression of the distributive homogeneity of the number of biological species and the number of individuals between varieties characterized by the Shannon-Wiener biodiversity index.

8-17 沿海地区风暴潮灾害情况

Survey of Storm Surges by Coastal Regions

受灾地区 Disaster Area	受灾人口 （万人） Disaster-stricken Population （10 000 persons）	死亡人数* （人） Death Toll （person）	受灾面积 Disaster-stricken area 农田 （千公顷） Disaster-Affected farmland （1 000 hm²）	水产养殖 （千公顷） Affected Area of Mariculture （1 000 hm²）
合　计 Total	**786.5**	**7**	**7.1**	**81.2**
山　东 Shandong	–	0	0.0	0.0
江　苏 Jiangsu	5.3	0	3.2	0.0
上　海 Shanghai	–	0	0.6	0.0
浙　江 Zhejiang	327.9	0	0.0	27.4
福　建 Fujian	34.2	0	3.1	27.5
广　东 Guangdong	376.4	5	0.3	24.4
广　西 Guangxi	42.8	0	0.0	1.6
海　南 Hainan	–	2	0.0	0.2

注："–"表示未统计。

Notes: "–" indicates no statistics.

8-17 续表 continued

受灾地区 Disaster Area	海岸工程 （千米） Coastal Engineering （km）	房屋 House （间）	船只 （艘） Boats	直接经济损失 （亿元） Direct Economic Loss （100 million yuan）
合 计 Total	**177.6**	**4 251**	**7 235**	**72.6**
山 东 Shandong	14.0	0	3	0.4
江 苏 Jiangsu	0.0	127	0	0.6
上 海 Shanghai	0.0	0	2	0.1
浙 江 Zhejiang	41.3	116	639	11.2
福 建 Fujian	88.7	3 858	4 265	30.8
广 东 Guangdong	29.0	78	2 325	28.8
广 西 Guangxi	4.5	72	0	0.5
海 南 Hainan	0.1	0	1	0.3

注：*包括失踪人数。

Notes: * includes missing people.

8-18 沿海地区赤潮灾害情况
Survey of Red Tide by Coastal Regions

时间 Date	影响区域 Area	最大面积 （平方千米） Disaster Area(km^2)
合 计 **Total**		**2 289**
1月23日至2月13日 Jan.23–Feb.13	儋州白马井和排浦镇近岸海域 Near Baimajing and Paipu town Sea Area of Zhanzhou	100
4月26日至5月3日 Apr.26–May.3	渔山列岛附近海域 Sea Area of Yushan Islands	200
5月20日至8月13日 May.20–Aug.13	绥中-滦河口海域 Sea Area of Suizhong–Luanhekou	825
5月26日至6月2日 May.26–Jun.2	霞浦县古镇海域 Guzhen Sea Area of Xiapu County	100
6月12日至21日 Jun.12–Jun.21	温州南麓海域 Nanlu Island Sea Area of Wenzhou	390
6月14日至16日 Jun.14–Jun.16	辽东湾西部海域 West Sea Area of Liaodong Bay	260
8月21日至9月3日 Aug.21–Sep.3	天津港南侧海域 West Sea Area of Tianjin Port	264
9月10日至19日 Sep.10–Sep.19	泉州安海湾、围头湾海域 Anhai and Weitou Bay Sea Area of Quanzhou	150

注：本表仅列出最大面积超过100平方千米（含）的赤潮过程。

Note: This table only lists the red tide processes with the maximum area more than 100 km^2.

主要统计指标解释

1. 工业废水排放量 指经过企业厂区所有排放口排到企业外部的工业废水量。包括生产废水、外排的直接冷却水、超标排放的矿井地下水和与工业废水混排的厂区生活污水，不包括外排的间接冷却水(清污不分流的间接冷却水应计算在内)。

2. 直接排入海的工业废水量 指经企业位于海边的排放口，直接排入海中的废水量。直接排入是指废水经过工厂的排污口直接排入海，而未经过城市下水道或其他中间体，也不受其他水体的影响。

3. 工业废水处理量 指报告期内各种水治理设施实际处理的工业废水量，包括处理后外排的和处理后回用的工业废水量，虽经处理但未达到国家或地方排放标准的废水量也应计算在内。计算时，如遇车间和厂排放口均有治理设施，并对同一废水分级处理时，不应重复计算工业废水处理量。

4. 一般工业固体废物处置量 指将固体废物焚烧或者最终置于符合环境保护规定要求的场所并不再回取的工业固体废物量(包括当年处置往年的工业固体废物累计贮存量)。

5. 一般工业固体废物倾倒丢弃量 指将所产生的固体废物倾倒丢弃到固体废物污染防治设施、场所以外的量。不包括矿山开采的剥离废石和掘进废石(煤矸石和呈酸性或碱性的废石除外)。

6. 当年安排施工项目数 指报告期内由国家、部门、地方或企业单位安排开工的，并以治理废水、废气、固体废物、噪声和其他(如电磁波、恶臭等)环境污染的环境治理工程的总数。不包括"三同时"项目。

7. 当年竣工项目数 指报告期内竣工投入运行的治理废水、废气、固体废物、噪声及其他污染的环境工程项目的总数。

Explanatory Notes on Main Statistical Indicators

1. Volume of Industrial Waste Water Discharged refers to the quantity of industrial waste water discharged externally through all the outlets in the factory area of the enterprise, including the waste water from production, externally discharged direct cooling water, mine -shaft groundwater discharged exceeding the set standard and the domestic sewage of the factory area discharged together with the industrial waste water, but not including the indirect cooling water discharge externally.

2. Volume of Industrial Waste Water Discharged Directly to Sea refers to the quantity of waste water directly discharged into the sea through the outlets of the enterprise by the sea. Direct discharge means the direct discharge into the sea of waste water through the outlets of the factory, which is not discharged via the urban sewers or other intermediates and is not affected by other water bodies.

3. Volume of Industrial Waste Water Treated refers to the industrial waste water volume actually treated by various water treatment facilities in the period covered by the report, including the quantity of the industrial waste water discharged and reused after treatment. The amount of the waste water which is not up to the state or local standard of discharge upon treatment should be included. If the workshops and outlets of the factory are provided with treatment facilities and carry out graded treatment of the same waste water, the processed volume of industrial waste water cannot be calculated repeatedly.

4. Volume of Common Industrial Solid Wastes Discharged refers to the amount of solid wastes discharged out of the facilities and sites for the solid wastes pollution prevention and control, not including the stripped and tunneled waste ores in the excavation of mines (other than gangues and acid or alkaline waste ores).

5. Volume of Common Industrial Solid Wastes Treated refers to the volume of industrial solid wastes which are to be burned or finally placed at the sites in keeping with the requirement of environmental protection and will not be recovered (including the accumulated amount of storage in former years of industrial solid radioactive matter disposed of in the year).

6. Arranged for Construction in the Year refers to the total number of environmental pollution control projects for controlling waste water, waste gas, solid wastes, noise and other environmental pollutions (such as electromagnetic wave, offensive odor) started by the state, governmental departments, local governments or enterprises in the period covered by the report mainly for the purpose of pollution control and multipurpose utilization of "three wastes".

7. Number of Pollution Treatment Projects Completed within the Current Year refers to the total number of environmental engineering projects for controlling waste water, waste gas, solid wastes, noise and other pollutions which are completed and put into operation in the period covered by the report.

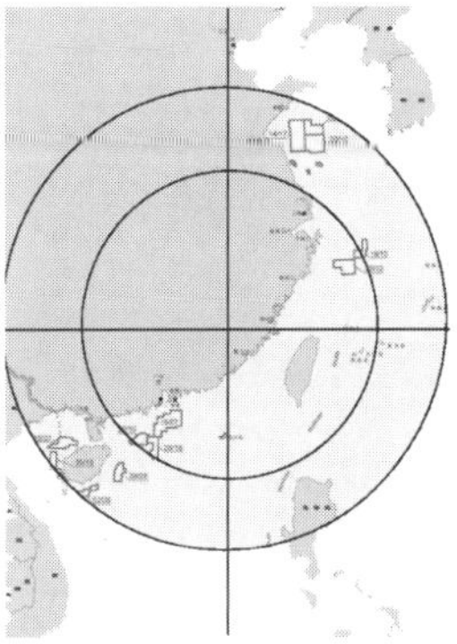

9
海洋行政管理及公益服务
Marine Administration and Public-Good Service

9-1　海域使用管理情况
Sea Area Use Management

地　区 Region	颁发海域使用权证书(本) Certificates of Right of Sea Area Use Issued (volume)	确权海域面积(公顷) Area of Waters with Established Rights (hm^2)	征收海域使用金(万元) Charge for Sea Area Utilization (10 000 yuan)
全国总计 National Total	**3 606**	**253 613.1**	**820 625.4**
天　津 Tianjin	37	1 208.6	101 815.7
河　北 Hebei	558	19 233.6	36 796.7
辽　宁 Liaoning	1 098	101 237.1	103 139.1
上　海 Shanghai	5	568.4	1 677.7
江　苏 Jiangsu	218	33 466.5	68 423.0
浙　江 Zhejiang	247	4 279.5	112 005.6
福　建 Fujian	288	6 702.9	84 573.3
山　东 Shandong	680	74 743.2	125 401.4
广　东 Guangdong	161	4 252.2	55 332.1
广　西 Guangxi	276	6 088.7	31 595.3
海　南 Hainan	33	1 693.3	91 181.0
其　他 Others	5	139.0	8 684.5

注:其他为沿海省(自治区、直辖市)管理海域以外(指渤海中部海域)。

Note: Others are the sea areas outside the control of coastal provinces, autonomous regions and municipalities directly under the Central Government. (Referring to the mid-Bohai Sea area)

9-2 依法批准无居民海岛开发利用情况
Development and Utilization of Uninhabited Islands Legally Approved

地区 Region	海岛名称 Island Name	用途 Use	用岛面积(公顷) Island Area(hm^2)	批准年份 Approved Year
辽宁 Liaoning	大笔架山 Dàbǐjià Shān	旅游娱乐 Tourist Recreation	3.2	2011
	西沙坨子岛 Xīshātuózi Dǎo	渔业开发 Fishery Development	1.4	2013
	空坨子 Kōng Tuózi	渔业开发 Fishery Development	0.1	2015
河北 Hebei	祥云岛 Xiángyún Dǎo	旅游娱乐 Tourist Recreation	1 492.8	2011
山东 Shandong	桃花岛 Táohuā Dǎo	旅游娱乐 Tourist Recreation	0.4	2012
浙江 Zhejiang	旦门山岛 Dànménshān Dǎo	旅游娱乐 Tourist Recreation	101.8	2011
	大羊屿 Dàyáng Yǔ	旅游娱乐 Tourist Recreation	26.5	2013
	扁鳗屿 Biǎnmàn Yǔ	公益服务 Public Service	0.2	2015
福建 Fujian	箭屿 Jiàn Yǔ	旅游娱乐 Tourist Recreation	1.5	2013
	小岁屿 Xiǎosuì Yǔ	交通运输 Transportation	8.7	2012
	连江洋屿 Liánjiāng Yángyǔ	旅游娱乐 Tourist Recreation	8.4	2012
	东埔石岛 Dōngpǔ Shídǎo	工业仓储 Industrial Warehousing	5.1	2015
广东 Guangdong	大三洲 Dàsān Zhōu	旅游娱乐 Tourist Recreation	1.5	2013
	小三洲 Xiǎosān Zhōu	旅游娱乐 Tourist Recreation	1.7	2013
广西 Guangxi	大娥眉岭 Dà'éméi Lǐng	交通运输 Transportation	1.0	2011
海南 Hainan	东锣岛 Dōngluó Dǎo	旅游娱乐 Tourist Recreation	11.4	2012

9-3 海洋倾废管理情况
Management on the Ocean Dumping of Wastes

海 区 Sea Area	签发疏浚物海洋倾倒许可证(份) Permit Issued for Ocean Dumping of Dredged Materials (unit)	新选划倾倒区数(个) Number of Newly Designated Dumping Zones (unit)
合 计 Total	**100**	**12**
渤黄海 Bohai Sea and Yellow Sea	23	0
东 海 East China Sea	38	12
南 海 South China Sea	39	0

9-4 海洋行政执法情况

Marine Administrative Enforcement

执法领域 Law Enforcement Area	检查项目（个） Number of Items Subject to Inspection (number)	检查次数（次） Times of Inspection (time)	发现违法行为（起） Illegal Acts Found (case)	作出行政处罚（件） Inflict Administrative Punishments (case)	决定罚款额（万元） Determine Fines (10 000 yuan)	已收缴罚款额（万元） The fine has been Seized (10 000 yuan)
合　计 Total	**12 601**	**27 686**	**75**	**82**	**444 070.9**	**401 832.5**
海域使用 Sea Area Use	3 133	6 504	33	22	437 285.8	395 217.6
涉外海洋科研 Foreign-related Marine Scientific Research	64	70				
海底电缆管道 Submarine Cable Protection	13	265			199.2	199.2
海洋工程建设项目环境保护 Environmental Protection for the Marine Engineering Construction Projects	2 086	5 370	17	24	3 125.3	3 034.6
海洋倾废 Oceanic Dumping of Wastes	773	4 819	23	33	596.2	595.7
海洋生态保护 Marine Ecological Conservation	67	333		2	2 534.3	2 533.3
海岛保护 Island Protection	6 465	10 325	2	1	330.1	252.1

9-5 沿海地区海滨观测台站分布概况
Distribution of Coastal Observation Stations by Coastal Regions

单位:个 (unit)

地 区 Region	合 计 Total	海洋站 Marine Station	验潮站① Tide Station	气象台站② Meteorological Station	地震台站 Seismic Station	雷达站 Radar Station
合 计 Total	**1 353**	**134**	**252**	**536**	**282**	**149**
天 津 Tianjin	25	1		14	7	3
河 北 Hebei	37	6		3	21	7
辽 宁 Liaoning	170	9	3	115	29	14
上 海 Shanghai	148	4	63	56	10	15
江 苏 Jiangsu③	94	11	26	34	18	5
浙 江 Zhejiang	158	24	37	42	21	34
福 建 Fujian	227	25	11	123	47	21
山 东 Shandong	151	20	22	52	39	18
广 东 Guangdong	213	20	82	32	56	23
广 西 Guangxi	41	5	5	19	8	4
海 南 Hainan	89	9	3	46	26	5

注: ①潮流量观测站44处,潮水位观测站208处。
②中国气象局所属的气象台站仅指沿海气象台站中有海洋气象观测、预报及服务等业务的台站,合计中含有中央气象台1个。
③江苏省的雷达站数据仅包括连云港市。

Notes: ①There are 44 tidal current observation stations and 208 tidal level observation stations.
② The meteorological observatories and stations under the China Meteorological Administration only refer to those with marine meteorological observation, forecast and service among the coastal meteorological observatories and the total includes the central meteorological station.
③The data of Jiangsu´s radar station only include those of the Lianyungang city.

9-6 海洋预报服务概况
Marine Forecast Service

单位:次 (time)

预报项目 Item	数值预报 Numerical Forecast			
	预报服务次数 Frequency	发布次数 Frequency of Release		
		广播电视 Radio and TV	互联网 Internet	纸质 Paper Media
合 计 Total	**35 412**	**1 835**	**33 237**	**242**
海 浪 Sea Wave	8 760	730	7 665	
海 温 Sea Surface Temperature	6 935	365	6 935	
潮 汐 Tide	6 388	730	5 840	
海 流 Sea Current	7 300		7 300	
海平面 Sea Level				
盐 度 Salinity	2 555		2 555	
赤 潮 Red Tide				
滨海旅游 Coastal Tourism	365		365	
海 冰 Sea Ice	96	10	90	6
绿 潮 Green Seaweed	195		135	60
溢 油 Oil Spill	28			28
厄尔尼诺 El Niño	12			
专 项 Special Item	1 476		1 460	16
其 他 Others	1 302		892	132

9-6 续表 continued

预报项目 Item	统计预报 Statistical Forecast			
	预报服务次数 Frequency	发布次数 Frequency of Release		
		广播电视 Radio and TV	互联网 Internet	纸 质 Paper Media
合 计 Total	**133 328**	**46 249**	**70 058**	**37 644**
海 浪 Sea Wave	42 097	14 158	21 135	10 820
海 温 Sea Surface Temperature	37 380	16 140	20 642	8 160
潮 汐 Tide	30 652	12 515	14 050	7 827
海 流 Sea Current				
海平面 Sea Level				
盐 度 Salinity				
赤 潮 Red Tide	484	158	222	314
滨海旅游 Coastal Tourism	3 835	2 100	3 013	1 279
海 冰 Sea Ice	2 212	83	391	1 798
绿 潮 Green Seaweed	44			44
溢 油 Oil Spill				
厄尔尼诺 El Niño	1 460		1 460	
专 项 Special Item	13 603	730	7 953	7 398
其 他 Others	1 561	365	1 192	4

注:本表只包含国家海洋局资料。

Note: This table only contains the data from the State Oceanic Administration.

9-7 海洋环境观测情况
Condition of Marine Environmental Observation

项目 Item	台站观测 Station Observation	断面观测 Sectional Observation	浮标观测 Buoy Observation	船舶测报 Ship Measuring and Reporting	其他观测 Other Observation
测站(点)数(个) Number of Stations (unit)	107	212	42	41	136
实际获得数据量(个) Quantity of Data Actually Obtained (unit)	326 029 902	512 230	6 093 374	22 674 879	8 390 524

9-8 海洋调查概况
Marine Survey Statistics

调查名称 Name	站点数(个) Number of Stations	船舶数(艘) Number of Ships	项目数(个) Number of Items	实际获得数据(个) Quantity of Data Actually Obtained	发布通(公、简)报量(期) Quantity of Circulars (Bulletin, Brief Reports)
合　计 Total	**2 908**	**627**	**785**	**324 492**	**10**
大洋调查 Oceanic Survey	393	9	45	103 062	
极地调查 Polar Survey	187	2	22	3 278	
专项调查 Special Survey	845	118	52	48 525	
其他调查 Other Surveys	1 483	498	666	169 627	10

9-9 涉外海洋科学研究审批情况
Examination and Approval of Foreign-Related Marine Scientific Research

单位:份 (unit)

审批单位 Examination and Approval Authority	研究活动申请审批 Examination and Approval of Application of Research Activities	船只作业计划审批 Examination and Approval of the Ship Operating Plan
国家海洋局 State Oceanic Administration, People's Republic of China	2	2

9-10 海洋档案及利用情况
Marine File and Its Use

指 标 Item	指 标 值 Data
室(馆)存档案 Files Deposited in the Archives	
纸介质档案(卷、册) Paper Media Files (reel,volume)	152 938
电子档案(GB) Electronic Media Files (GB)	37 108
本年接收档案 Files Received this Year	
纸质档案(卷、册) Paper Media Files (reel,volume)	25 013
电子档案(GB) Electronic Media Files (GB)	12 994
本年利用档案 Files Utilized this Year	
纸质档案(件) Paper Media Files (piece)	12 417
电子档案(GB) Electronic Media Files (GB)	62

9-11 卫星遥感接收应用情况
Remote-Sensing Receiving and Utilization

指　标 Item	指 标 值 Data
卫星接收次数(轨) Number of Satellite Receptions (orbit / track)	92 195
全年接收时间(天) Whole Year Receiving Time (day)	3 245
全年实际接收存档数据量(GB) Amount of Data Actually Received and Placed on File in the Whole Year (GB)	112 548.1
累计存档数据量(GB) Total Amount of Data Placed on File (GB)	80 884.5
卫星数据分发 Satellite Data Distribution	
类别用户(个) Classified Users (unit)	8
分发数据量(GB) Amount of Data Distributed (GB)	100 407.7

9-12 海洋标准化监督管理情况
Supervision and Management of Marine Standardization

单位:项 （item）

指 标 Item	指 标 值 Data
标准立项审查 Examination of the Standards for Authorization	94
国家标准 National Standards	22
行业标准 Professional Standards	72
标准审查 Examination of Standards	270
国家标准 National Standards	53
行业标准 Professional Standards	173
标准出版 Standards Publication	28
国家标准 National Standards	7
行业标准 Professional Standards	21
标准实施监督检查(次) Supervision and Examination of Standards Implementation （time）	3

主要统计指标解释

1. 海域使用检查 针对不同类型的用海行为进行的监督检查。

2. 涉外海洋科研项目检查 主要针对国际组织、外国组织和个人为和平目的，单独或者与中华人民共和国的组织合作，使用船舶或者其他运载工具、设施，在中华人民共和国内海、领海以及中华人民共和国管辖的其他海域内进行的对海洋环境和海洋资源等的调查研究活动进行监督检查。

3. 海底电缆管道检查 主要是针对铺设海底电缆管道路由调查、铺设施工和维修改造等的监督检查。

4. 海洋工程建设项目环境保护检查 主要是针对防治海洋工程建设项目对海洋环境的污染损害的监督检查。

5. 海洋倾废检查 主要是针对防治倾倒废弃物对海洋环境的污染损害的监督检查。

6. 海洋生态保护检查 主要是针对红树林、珊瑚礁、滨海湿地、海岛、海湾、入海河口、重要渔业水域等具有典型性、代表性的海洋生态系统，珍稀、濒危海洋生物的天然集中分布区，具有重要经济价值的海洋生物生存区域及有重大科学文化价值的海洋自然历史遗迹和自然景观等海洋自然保护区以及其他需要予以特殊保护的区域的监督检查。

7. 断面观测 按照国家海洋局“断面监测方案”，每年定期利用船舶在沿海设定的断面上进行海洋水文、气象、生物、化学等项目的监测活动。

8. 浮标观测 在海上固定站位获取长期、连续海洋环境观测资料的海上锚定资料浮标。

9. 船舶测报 利用在固定航线上走航的商船或渔船，每日定时所在地的海洋环境状况（主要是水文气象要素），并将监测数据实时发送给有关单位，供海洋环境预报使用。

10. 大洋调查 以大洋科考、研究为目的的远洋调查。

11. 专项调查 为完成国家专项任务进行的海洋调查。

12. 全年接收时间 指全年整个应用系统接收的各类遥感卫星数据时，接收设备工作时间。

13. 全年实际接收存档数据量 指全年整个应用系统接收的各类遥感卫星数据原始数据量。

14. 累计存档数据量 指全年整个应用系统制作的各类遥感产品数据量。

15. 分发数据量 指应用系统提供数据产品用于开展各类应用的数据量。

16. 卫星接收次数 地面观测系统接收卫星数据的数量。

17. 国家标准 针对海洋领域内需要在全国范围内统一的有关技术要求所制定的国家标准。海洋国家标准由国家标准化主管部门统一批准、编号和发布。

18. 行业标准 对没有海洋国家标准而又需要在海洋领域内统一的技术要求所制定的标准。海洋行业标准由国家海洋局统一批准、编号和发布。

Explanatory Notes on Main Statistical Indicators

1. Sea Area Use Supervision and Inspection refers to the various types of sea area use conducts.

2. Inspection of Foreign–Related Marine Scientific Research Projects refers to the supervision and inspection of the activities of surveying marine environment and resources conducted by international organizations, foreign organizations and individuals for peaceful purposes, alone or in cooperation with PRC organizations, by using ships or other means of delivery as facilities in the internal seas and territorial waters of the People's Republic of China as well as in the other water under the jurisdiction of the People's Republic of China.

3. Inspection of Submarine Cables and Pipelines is mainly aimed at the survey of the submarine cables and pipelines laying route and supervision and inspection of the laying construction, repair and transformation.

4. Environmental Protection Inspection for the Marine Engineering Construction Projects is mainly directed against preventing and controlling the pollution damage of the marine engineering construction project to the marine environment.

5. Inspection of Oceanic Dumping of Wastes mainly refers to the supervision and inspection aimed at preventing and controlling the pollution damage of wastes dumping to the marine environment.

6. Inspection of Marine Ecological Protection mainly refers to the supervision and inspection of the typical and representative marine ecosystems such as mangroves, coral reef, coastal wetland, sea island, bay, estuaries open to the sea, important fishery waters, etc, the natural centralized distribution zones of rare and endangered marine life, the living areas for the marine life with significant economic values and the marine nature reserves such as the marine natural and historical remains and natural landscapes with important scientific and cultural values as well as other areas needing to be specially protected.

7. Sectional Monitoring According to the "Sectional Monitoring Plan" of the State Oceanic Administration, monitoring activities concerning such items as marine hydrology, meteorology, biology and chemistry are carried out regularly every year on the sections set in the coastal area.

8. Buoy Monitoring refers to the monitoring carried out by the offshore mooring data buoys which acquire long–term, continuous marine environmental observations at the fixed stations at sea.

9. Ship Monitoring Monitoring the marine environmental condition in the sea area of fixed times every day by using the merchant or fishing vessels cruising on the fixed navigation line (mainly the hydro meteorological elements) and transmitting the monitored data in real time to the related units for use in the marine environmental forecast.

10. Oceanic Survey refers to the oceanic surveys aimed at the oceanic scientific investigations and research.

11. Special–Subject Survey refers to the oceanic investigation for the purpose of fulfilling the state's special tasks.

12. Whole Year Receiving Time refers to the operating time of receiving equipment in the whole year when the whole application system receives all types of remote sensing satellite data.

13. Amount of Data Actually Received and Placed on File throughout the year refers to the raw data of remote–sensing satellite data of various kinds received by the whole application system throughout the year.

14. Total Amount of Date Placed on File refers to the date amount of various remote–sensing products made by the whole application systems throughout the year.

15. Amount of Data Distributed refers to the amount of data products provided by the application system for various uses.

16. Number of Satellite Receptions refers to the amount of data received by the ground observation system.

17. National Standards refers to the standards formulated in view of the relevant technical requirements in the marine field that need to be unified throughout the country. The Marine National standards are approved, numbered and issued uniformly by the state department responsible for standardization.

18. Professional Standards refers to the standards formulated for the technical requirements which have no national standards but need to be unified in the marine field. The Marine Professional Standards are approved, numbered and issued by the State Oceanic Administration.

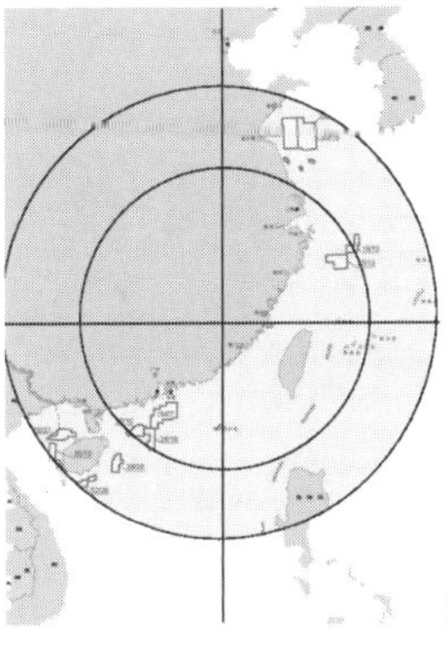

10

全国及沿海社会经济

National and Coastal Socioeconomy

10-1 国内生产总值
Gross Domestic Product

单位:亿元 (100 million yuan)

年 份 Year	国内生产总值 Gross Domestic Product	第一产业 Primary Industry	第二产业 Secondary Industry	第三产业 Tertiary Industry
2001	110 863.1	15 502.5	49 660.7	45 700.0
2002	121 717.4	16 190.2	54 105.5	51 421.7
2003	137 422.0	16 970.2	62 697.4	57 754.4
2004	161 840.2	20 904.3	74 286.9	66 648.9
2005	187 318.9	21 806.7	88 084.4	77 427.8
2006	219 438.5	23 317.0	104 361.8	91 759.7
2007	270 232.3	27 788.0	126 633.6	115 810.7
2008	319 515.5	32 753.2	149 956.6	136 805.8
2009	349 081.4	34 161.8	160 171.7	154 747.9
2010	413 030.3	39 362.6	191 629.8	182 038.0
2011	489 300.6	46 163.1	227 038.8	216 098.6
2012	540 367.4	50 902.3	244 643.3	244 821.9
2013	595 244.4	55 329.1	261 956.1	277 959.3
2014	643 974.0	58 343.5	277 571.8	308 058.6
2015	685 505.8	60 870.5	280 560.3	344 075.0

10-2 国内生产总值增长速度
Growth Rate of Gross Domestic Product

单位:% (%)

年 份 Year	国内生产总值 Gross Domestic Product	第一产业 Primary Industry	第二产业 Secondary Industry	第三产业 Tertiary Industry
2001	8.3	2.6	8.5	10.3
2002	9.1	2.7	9.9	10.5
2003	10.0	2.4	12.7	9.5
2004	10.1	6.1	11.1	10.1
2005	11.4	5.1	12.1	12.4
2006	12.7	4.8	13.5	14.1
2007	14.2	3.5	15.1	16.1
2008	9.7	5.2	9.8	10.5
2009	9.4	4.0	10.3	9.6
2010	10.6	4.3	12.7	9.7
2011	9.5	4.2	10.7	9.5
2012	7.9	4.5	8.4	8.0
2013	7.8	3.8	8.0	8.3
2014	7.3	4.1	7.4	7.8
2015	6.9	3.9	6.1	8.3

注:本表按可比价格计算(上年为基期)。

Note:This table is calculated at the comparable price (with the previous year as the base period).

10-3 沿海地区生产总值
Gross Regional Product of Coastal Regions

单位:亿元 (100 million yuan)

年 份 Year	地区生产总值 Gross Regional Product	第一产业 Primary Industry	第二产业 Secondary Industry	第三产业 Tertiary Industry
合 计 Total	**395 440.2**	**25 824.0**	**178 637.8**	**190 978.5**
天 津 Tianjin	16 538.2	208.8	7 704.2	8 625.2
河 北 Hebei	29 806.1	3 439.5	14 386.9	11 979.8
辽 宁 Liaoning	28 669.0	2 384.0	13 042.0	13 243.0
上 海 Shanghai	25 123.5	109.8	7 991.0	17 022.6
江 苏 Jiangsu	70 116.4	3 986.1	32 044.5	34 085.9
浙 江 Zhejiang	42 886.5	1 832.9	19 711.7	21 341.9
福 建 Fujian	25 979.8	2 118.1	13 064.8	10 796.9
山 东 Shandong	63 002.3	4 979.1	29 485.9	28 537.4
广 东 Guangdong	72 812.6	3 345.5	32 613.5	36 853.5
广 西 Guangxi	16 803.1	2 565.5	7 717.5	6 520.2
海 南 Hainan	3 702.8	854.7	875.8	1 972.2

10-4 沿海地区生产总值增长速度
Growth Rate of Gross Regional Product of Coastal Regions

单位:%　　　　(%)

地 区 Region	2006	2007	2008	2009	2010	2011	2012	2013	2014	2015
天 津 Tianjin	14.7	15.5	16.5	16.5	17.4	16.4	13.8	12.5	10.0	9.3
河 北 Hebei	13.4	12.8	10.1	10.0	12.2	11.3	9.6	8.2	6.5	6.8
辽 宁 Liaoning	14.2	15.0	13.4	13.1	14.2	12.2	9.5	8.7	5.8	3.0
上 海 Shanghai	12.7	15.2	9.7	8.2	10.3	8.2	7.5	7.7	7.0	6.9
江 苏 Jiangsu	14.9	14.9	12.7	12.4	12.7	11.0	10.1	9.6	8.7	8.5
浙 江 Zhejiang	13.9	14.7	10.1	8.9	11.9	9.0	8.0	8.2	7.6	8.0
福 建 Fujian	14.8	15.2	13.0	12.3	13.9	12.3	11.4	11.0	9.9	9.0
山 东 Shandong	14.7	14.2	12.0	12.2	12.3	10.9	9.8	9.6	8.7	8.0
广 东 Guangdong	14.8	14.9	10.4	9.7	12.4	10.0	8.2	8.5	7.8	8.0
广 西 Guangxi	13.6	15.1	12.8	13.9	14.2	12.3	11.3	10.2	8.5	8.1
海 南 Hainan	13.2	15.8	10.3	11.7	16.0	12.0	9.1	9.9	8.5	7.8

注:本表按可比价格计算(上年为基期)。

Note: This table is calculated at the comparable price (with the previous year as the base period).

10-5 沿海城市生产总值(2014年)

Gross Regional Product of Coastal Cities, 2014

单位:亿元 (100 million yuan)

沿海城市 Coastal City		地区生产总值 Gross Regional Product	第一产业 Primary Industry	第二产业 Secondary Industry	第三产业 Tertiary Industry
合　计	**Total**	**220 210.1**	**10 712.8**	**101 937.8**	**107 557.9**
天　津	**Tianjin**	**15 726.9**	**199.9**	**7 731.6**	**7 795.2**
河　北	**Hebei**	**10 558.7**	**1 050.7**	**5 672.8**	**3 834.5**
唐　山	Tangshan	6 225.3	558.4	3 595.1	2 071.2
秦皇岛	Qinhuangdao	1 200.0	174.6	449.3	576.1
沧　州	Cangzhou	3 133.4	317.7	1 628.4	1 187.2
辽　宁	**Liaoning**	**13 613.8**	**1 108.3**	**6 625.8**	**5 879.8**
大　连	Dalian	7 655.6	441.7	3 697.6	3 516.2
丹　东	Dandong	1 022.6	144.7	459.4	418.6
锦　州	Jinzhou	1 364.0	201.6	632.9	529.5
营　口	Yingkou	1 546.1	110.5	776.9	658.6
盘　锦	Panjin	1 304.0	114.6	739.3	450.0
葫芦岛	Huludao	721.6	95.1	319.7	306.9
上　海	**Shanghai**	**23 567.7**	**124.3**	**8 167.7**	**15 275.7**
江　苏	**Jiangsu**	**11 454.2**	**1 091.2**	**5 484.4**	**4 878.8**
南　通	Nantong	5 652.7	339.7	2 812.2	2 500.8
连云港	Lianyungang	1 965.9	262.1	889.8	814.3
盐　城	Yancheng	3 835.6	489.4	1 782.4	1 563.8
浙　江	**Zhejiang**	**33 140.6**	**1 323.4**	**15 886.8**	**15 929.7**
杭　州	Hangzhou	9 206.2	274.3	3 845.4	5 086.4
宁　波	Ningbo	7 610.3	275.5	3 980.2	3 353.9
温　州	Wenzhou	4 303.0	117.9	2 029.8	2 155.4
嘉　兴	Jiaxing	3 352.6	144.8	1 813.8	1 394.0
绍　兴	Shaoxing	4 265.9	194.1	2 213.6	1 858.2
舟　山	Zhoushan	1 015.3	100.9	425.3	489.1
台　州	Taizhou	3 387.4	215.8	1 578.9	1 592.8
福　建	**Fujian**	**19 560.6**	**1 310.7**	**10 185.6**	**8 064.5**
福　州	Fuzhou	5 169.2	416.1	2 352.0	2 401.1
厦　门	Xiamen	3 273.6	23.6	1 460.3	1 789.7
莆　田	Putian	1 502.1	109.8	866.7	525.4
泉　州	Quanzhou	5 733.4	172.6	3 553.5	2 007.8
漳　州	Zhangzhou	2 506.4	350.4	1 247.4	908.3
宁　德	Ningde	1 376.1	238.2	705.7	432.2

10-5 续表 continued

沿海城市 Coastal City	地区生产总值 Gross Regional Product	第一产业 Primary Industry	第二产业 Secondary Industry	第三产业 Tertiary Industry
山 东 Shandong	**29 589.6**	**1 905.9**	**15 099.0**	**12 584.7**
青 岛 Qingdao	8 692.1	349.4	3 890.6	4 452.1
东 营 Dongying	3 430.5	116.0	2 284.7	1 029.8
烟 台 Yantai	6 002.1	426.8	3 179.9	2 395.4
潍 坊 Weifang	4 786.0	456.1	2 431.8	1 898.1
威 海 Weihai	2 790.3	207.6	1 355.0	1 228.0
日 照 Rizhao	1 611.9	139.3	811.9	660.7
滨 州 Binzhou	2 276.7	210.8	1 145.2	920.5
广 东 Guangdong	**59 204.1**	**2 070.1**	**25 647.2**	**31 486.4**
广 州 Guangzhou	16 706.9	218.9	5 591.8	10 897.9
深 圳 Shenzhen	16 001.8	4.8	6 812.0	9 183.4
珠 海 Zhuhai	1 867.2	43.9	938.7	884.5
汕 头 Shantou	1 716.5	91.8	903.6	721.1
江 门 Jiangmen	2 082.8	168.1	1 021.6	893.1
湛 江 Zhanjiang	2 259.0	429.4	894.1	935.5
茂 名 Maoming	2 349.0	364.1	974.9	1 010.1
惠 州 Huizhou	3 000.4	141.0	1 697.0	1 162.3
汕 尾 Shanwei	717.0	109.8	333.2	274.0
阳 江 Yangjiang	1 168.5	192.9	561.3	414.3
东 莞 Dongguan	5 881.3	20.6	2 794.2	3 066.5
中 山 Zhongshan	2 823.0	66.9	1 560.8	1 195.3
潮 州 Chaozhou	850.2	60.8	466.2	323.3
揭 阳 Jieyang	1 780.4	157.0	1 098.0	525.2
广 西 Guangxi	**2 299.9**	**416.1**	**1 133.8**	**750.0**
北 海 Beihai	856.0	151.3	454.5	250.1
防城港 Fangchenggang	588.9	70.9	340.4	177.7
钦 州 Qinzhou	855.0	193.9	338.9	322.2
海 南 Hainan	**1 494.0**	**112.3**	**303.0**	**1 078.7**
海 口 Haikou	1 091.7	57.1	217.5	817.1
三 亚 Sanya	402.3	55.2	85.6	261.5

注:本表各省数据为合计数。

Note: The data for the provinces are the totals.

10-6 县级单位主要统计指标
Main Indicators of Regions at County Level

单位:亿元 （100 million yuan）

沿海县 Coastal County		第一产业增加值 Value-added of Primary Industry		第二产业增加值 Value-added of Secondary Industry	
		2014	2015	2014	2015
合　计	**Total**	**53 681 181**	**55 888 202**	**237 682 025**	**240 968 616**
河　北	**Hebei**	**3 487 284**	**3 476 907**	**8 770 212**	**7 872 780**
丰　南	Fengnan	423 336	471 462	4 129 243	3 714 337
滦　南	Luannan	832 163	856 425	1 094 262	944 010
乐　亭	Laoting	799 764	863 597	1 049 363	985 083
昌　黎	Changli	633 151	556 470	782 162	734 255
抚　宁	Funing	450 530	333 399	496 288	280 367
黄　骅	Huanghua	276 511	324 994	1 051 158	1 028 216
海　兴	Haixing	71 829	70 560	167 736	186 512
辽　宁	**Liaoning**	**6 863 883**	**7 193 167**	**22 102 760**	**19 781 223**
长　海	Changhai	504 888	470 014	72 092	72 123
瓦房店	Wafangdian	902 108	943 210	6 729 380	5 909 195
普兰店	Pulandian	916 613	980 254	4 258 829	3 921 754
庄　河	Zhuanghe	1 216 647	1 270 383	3 412 284	3 078 274
东　港	Donggang	715 405	775 879	1 583 145	1 325 368
凌　海	Linghai	545 290	552 864	1 548 569	1 331 563
盖　州	Gaizhou	374 047	377 750	919 607	775 337
大　洼	Dawa	639 749	676 612	1 756 472	1 793 923
盘　山	Panshan	470 944	498 065	690 885	676 643
绥　中	Suizhong	393 418	449 281	619 881	465 476
兴　城	Xingcheng	184 774	198 855	511 616	431 567
江　苏	**Jiangsu**	**7 769 267**	**8 138 946**	**28 924 504**	**30 770 161**
海　安	Hai′an	512 687	537 710	3 041 312	3 231 845
如　东	Rudong	621 674	649 665	2 971 100	3 148 317
启　东	Qidong	629 331	654 460	3 690 256	3 891 337

10-6 续表1 continued

沿海县 Coastal County		公共财政收入 Public Revenue of Local Governments		公共财政支出 Public Expenditure of Local Governments	
		2014	2015	2014	2015
合　计	**Total**	**36 273 543**	**37 625 074**	**50 637 446**	**58 749 872**
河　北	**Hebei**	**754 202**	**831 030**	**1 538 601**	**1 887 392**
丰　南	Fengnan	258 316	290 948	359 025	410 245
滦　南	Luannan	95 000	95 589	226 999	254 088
乐　亭	Laoting	100 117	111 160	213 035	288 112
昌　黎	Changli	81 383	91 359	211 209	249 471
抚　宁	Funing	74 838	81 506	181 787	203 111
黄　骅	Huanghua	117 705	130 088	246 113	344 852
海　兴	Haixing	26 843	30 380	100 433	137 513
辽　宁	**Liaoning**	**3 319 447**	**1 742 812**	**4 714 433**	**3 999 912**
长　海	Changhai	48 123	43 609	83 216	89 664
瓦房店	Wafangdian	728 145	376 450	806 611	621 085
普兰店	Pulandian	433 472	224 795	517 878	379 886
庄　河	Zhuanghe	418 251	229 103	583 670	449 470
东　港	Donggang	335 486	132 195	510 661	471 413
凌　海	Linghai	210 246	85 678	344 636	308 553
盖　州	Gaizhou	110 294	101 239	305 282	306 272
大　洼	Dawa	401 451	263 060	583 565	456 052
盘　山	Panshan	348 824	76 486	370 158	286 858
绥　中	Suizhong	150 155	90 532	320 000	312 446
兴　城	Xingcheng	135 000	119 665	288 756	318 213
江　苏	**Jiangsu**	**5 881 189**	**6 721 871**	**8 478 794**	**9 931 391**
海　安	Hai´an	541 022	620 566	705 844	843 989
如　东	Rudong	500 127	585 447	791 961	938 165
启　东	Qidong	672 463	768 609	740 076	862 358

10-6 续表2 continued

沿海县 Coastal County		第一产业增加值 Value-added of Primary Industry		第二产业增加值 Value-added of Secondary Industry	
		2014	2015	2014	2015
海 门	Haimen	490 663	517 975	4 431 636	4 715 945
赣 榆	Ganyu	738 037	699 700	2 121 900	2 302 700
东 海	Donghai	589 420	618 900	1 631 300	1 737 100
灌 云	Guanyun	551 289	598 100	1 255 000	1 335 200
灌 南	Guannan	447 523	482 300	1 287 600	1 368 600
响 水	Xiangshui	388 050	403 500	1 051 100	1 136 752
滨 海	Binhai	538 755	559 741	1 386 300	1 489 758
射 阳	Sheyang	731 092	784 100	1 394 700	1 476 102
东 台	Dongtai	828 646	879 195	2 618 100	2 792 105
大 丰	Dafeng	702 100	753 600	2 044 200	2 144 400
浙 江	**Zhejiang**	**6 298 914**	**6 606 480**	**55 669 748**	**56 675 069**
象 山	Xiangshan	579 451	613 666	1 806 461	1 846 924
宁 海	Ninghai	393 016	411 164	2 131 959	2 242 975
余 姚	Yuyao	413 400	423 578	4 677 312	4 683 538
慈 溪	Cixi	484 849	497 096	6 379 252	6 560 527
奉 化	Fenghua	285 034	290 288	1 405 983	1 432 944
洞 头	Dongtou	42 100	54 000	206 700	290 000
平 阳	Pingyang	134 546	144 839	1 568 716	1 458 333
苍 南	Cangnan	272 537	296 232	1 696 965	1 754 813
瑞 安	Rui'an	189 618	206 440	3 226 595	3 290 000
乐 清	Yueqing	199 926	209 106	3 884 521	3 909 661
海 盐	Haiyan	210 604	207 391	2 044 989	2 288 589
海 宁	Haining	227 531	215 675	3 798 051	3 844 508
平 湖	Pinghu	164 822	147 090	2 896 938	2 846 236
越 城	Yuecheng				
柯 桥	Keqiao	329 185	339 097	6 306 940	6 446 215
上 虞	Shangyu	413 559	426 260	3 720 940	3 909 580

10-6 续表3 continued

沿海县 Coastal County		公共财政收入 Public Revenue of Local Governments		公共财政支出 Public Expenditure of Local Governments	
		2014	2015	2014	2015
海门	Haimen	685 712	783 980	736 171	855 534
赣榆	Ganyu	401 689	445 078	664 898	755 798
东海	Donghai	372 008	410 163	617 361	687 167
灌云	Guanyun	355 892	393 199	557 402	618 031
灌南	Guannan	351 217	388 511	525 397	609 084
响水	Xiangshui	278 242	325 412	436 672	508 150
滨海	Binhai	334 513	386 540	580 745	694 488
射阳	Sheyang	175 000	205 588	450 670	593 838
东台	Dongtai	613 138	715 512	842 577	1 017 151
大丰	Dafeng	600 166	693 266	829 020	947 638
浙江	**Zhejiang**	**8 583 078**	**9 775 208**	**10 490 788**	**12 582 934**
象山	Xiangshan	331 678	379 485	527 444	613 051
宁海	Ninghai	360 643	415 712	509 391	602 441
余姚	Yuyao	648 070	750 989	804 673	925 207
慈溪	Cixi	1 000 151	1 122 589	1 078 856	1 254 760
奉化	Fenghua	290 233	338 192	476 853	607 125
洞头	Dongtou	45 021	54 172	163 861	187 916
平阳	Pingyang	220 002	256 688	405 822	457 164
苍南	Cangnan	256 366	320 667	494 925	544 021
瑞安	Rui´an	484 417	547 214	553 097	733 731
乐清	Yueqing	556 525	955 544	613 477	702 225
海盐	Haiyan	270 280	316 288	283 962	400 298
海宁	Haining	600 260	691 221	606 845	765 612
平湖	Pinghu	456 780	505 780	449 179	538 324
越城	Yuecheng				
柯桥	Keqiao	820 088	978 488	714 241	985 258
上虞	Shangyu	465 581	535 837	500 112	600 132

10-6 续表4 continued

沿海县 Coastal County		第一产业增加值 Value-added of Primary Industry		第二产业增加值 Value-added of Secondary Industry	
		2014	2015	2014	2015
岱 山	Daishan	291 326	324 500	999 869	1 065 014
嵊 泗	Shengsi	200 466	232 345	122 411	130 184
玉 环	Yuhuan	269 149	293 853	2 485 695	2 407 779
三 门	Sanmen	229 206	247 308	609 870	642 074
温 岭	Wenling	582 928	623 656	3 698 582	3 563 446
临 海	Linhai	385 661	402 896	2 000 999	2 061 729
福 建	**Fujian**	**7 550 588**	**8 034 107**	**46 270 569**	**44 880 248**
连 江	Lianjiang	1 109 635	1 199 308	1 312 150	1 391 522
罗 源	Luoyuan	306 094	336 836	1 140 537	1 120 198
平 潭	Pingtan	343 482	359 540	552 200	594 296
福 清	Fuqing	882 698	909 498	3 775 704	192 445
长 乐	Changle	427 178	432 828	3 592 249	3 731 900
仙 游	Xianyou	290 440	305 583	1 383 067	1 590 065
惠 安	Hui´an	278 549	289 695	4 654 900	5 025 210
金 门	Jinmen				
石 狮	Shishi	194 937	200 167	3 797 337	3 835 133
晋 江	Jinjiang	186 351	195 941	9 989 613	10 324 385
南 安	Nan´an	232 767	247 201	4 943 300	5 180 200
云 霄	Yunxiao	237 386	289 943	663 630	744 143
漳 浦	Zhangpu	600 945	622 951	1 445 208	1 291 365
诏 安	Zhao´an	356 475	392 790	742 071	835 472
东 山	Dongshan	277 883	308 588	677 356	742 825
龙 海	Longhai	558 234	575 369	3 257 882	3 636 115
霞 浦	Xiapu	476 248	516 629	531 575	563 927
福 安	Fu´an	413 169	438 556	2 082 277	2 214 758
福 鼎	Fuding	378 117	412 684	1 729 513	1 866 289

10-6 续表5 continued

沿海县 Coastal County		公共财政收入 Public Revenue of Local Governments		公共财政支出 Public Expenditure of Local Governments	
		2014	2015	2014	2015
岱 山	Daishan	114 443	126 037	298 684	355 707
嵊 泗	Shengsi	59 170	69 010	197 659	210 244
玉 环	Yuhuan	644 759	344 987	399 668	481 184
三 门	Sanmen	127 657	144 987	253 475	305 551
温 岭	Wenling	478 388	541 213	645 088	726 079
临 海	Linhai	352 566	380 108	513 476	586 904
福 建	**Fujian**	**6 939 470**	**6 582 710**	**8 169 395**	**9 224 606**
连 江	Lianjiang	334 948	381 306	453 798	460 637
罗 源	Luoyuan	135 766	108 899	239 158	201 685
平 潭	Pingtan	140 333	250 156	709 576	1 241 429
福 清	Fuqing	1 084 635	812 583	974 407	698 100
长 乐	Changle	314 854	336 452	361 596	473 385
仙 游	Xianyou	262 149	281 689	350 158	427 466
惠 安	Hui´an	308 702	330 203	477 517	545 591
金 门	Jinmen				
石 狮	Shishi	604 451	592 933	449 406	483 711
晋 江	Jinjiang	1 141 019	1 172 008	1 263 935	1 297 465
南 安	Nan´an	746 988	627 698	539 271	621 394
云 霄	Yunxiao	86 718	74 722	191 808	253 458
漳 浦	Zhangpu	294 217	315 538	369 780	480 195
诏 安	Zhao´an	81 413	80 294	206 119	271 449
东 山	Dongshan	173 949	182 712	197 483	277 020
龙 海	Longhai	706 284	480 409	519 883	442 963
霞 浦	Xiapu	109 146	116 707	218 444	274 111
福 安	Fu´an	227 896	235 026	352 041	418 564
福 鼎	Fuding	186 002	203 375	295 015	355 983

10-6 续表6 continued

沿海县 Coastal County		第一产业增加值 Value-added of Primary Industry		第二产业增加值 Value-added of Secondary Industry	
		2014	2015	2014	2015
山 东	**Shandong**	**7 265 326**	**8 642 227**	**8 753 490**	**10 689 294**
胶 州	Jiaozhou	492 947	532 000	4 972 800	5 178 000
即 墨	Jimo	592 823	601 922	5 187 100	5 980 400
胶 南	Jiaonan				
垦 利	Kenli	189 047	714 657	2 410 200	5 132 557
利 津	Lijin	266 796	270 753	1 322 458	1 326 837
广 饶	Guangrao	415 152	421 601	5 195 486	5 133 361
长 岛	Changdao	351 583	361 040	57 625	36 817
龙 口	Longkou	348 511	361 854	5 942 369	6 064 091
莱 阳	Laiyang	432 042	448 185	1 627 987	1 522 801
莱 州	Laizhou	667 172	681 153	3 649 423	3 734 380
蓬 莱	Penglai	275 768	280 713	2 610 451	2 485 597
招 远	Zhaoyuan	379 569	400 114	3 538 602	3 383 443
海 阳	Haiyang	602 496	629 594	1 389 194	1 045 117
寿 光	Shouguang	892 238	931 600	3 591 339	3 675 400
昌 邑	Changyi	334 906	347 350	1 976 000	2 040 400
文 登	Wendeng	550 520	580 521	3 356 564	3 329 730
荣 成	Rongcheng	805 360	831 341	4 803 159	4 747 443
乳 山	Rushan	375 266	395 467	2 305 684	2 266 586
无 棣	Wudi	360 197	361 158	1 271 130	1 283 785
沾 化	Zhanhua	372 896	387 020	615 197	617 515
广 东	**Guangdong**	**8 625 865**	**9 181 658**	**18 519 117**	**20 662 373**
南 澳	Nan´ao	37 676	39 438	52 970	55 816
台 山	Taishan	554 180	564 884	1 806 306	1 725 306
恩 平	Enping	189 710	200 362	468 192	515 142
遂 溪	Suixi	931 110	991 268	739 055	772 990
徐 闻	Xuwen	672 370	677 456	135 064	139 451
廉 江	Lianjiang	884 831	947 462	1 608 910	1 775 284

10-6 续表7 continued

沿海县 Coastal County		公共财政收入 Public Revenue of Local Governments		公共财政支出 Public Expenditure of Local Governments	
		2014	2015	2014	2015
山　东	**Shandong**	**8 705 289**	**9 538 043**	**55 822 768**	**58 984 260**
胶　州	Jiaozhou	675 066	800 616	788 072	929 001
即　墨	Jimo	790 400	930 493	887 600	1 148 887
胶　南	Jiaonan				
垦　利	Kenli	208 659	705 028	265 887	853 484
利　津	Lijin	117 619	125 000	221 295	232 814
广　饶	Guangrao	400 020	424 045	501 813	513 060
长　岛	Changdao	11 728	12 901	83 950	68 781
龙　口	Longkou	795 057	900 517	787 677	869 280
莱　阳	Laiyang	125 021	138 376	248 186	295 820
莱　州	Laizhou	520 678	572 757	556 912	619 232
蓬　莱	Penglai	272 058	300 017	331 586	376 966
招　远	Zhaoyuan	465 000	502 200	492 671	534 618
海　阳	Haiyang	247 146	269 567	288 331	333 557
寿　光	Shouguang	792 000	900 517	826 874	948 700
昌　邑	Changyi	251 969	273 326	311 556	421 786
文　登	Wendeng	438 903	492 468	516 353	579 255
荣　成	Rongcheng	600 105	672 789	814 389	1 006 177
乳　山	Rushan	276 588	310 358	350 315	406 036
无　棣	Wudi	174 129	197 752	287 132	332 118
沾　化	Zhanhua	103 180	113 500	192 891	219 722
广　东	**Guangdong**	**1 852 543**	**1 956 123**	**5 272 025**	**7 467 577**
南　澳	Nan´ao	18 858	21 408	95 302	89 637
台　山	Taishan	227 148	247 639	386 520	431 940
恩　平	Enping	95 283	98 898	201 020	248 023
遂　溪	Suixi	61 390	69 476	230 995	362 115
徐　闻	Xuwen	40 345	44 664	273 430	273 562
廉　江	Lianjiang	94 748	110 682	446 395	568 868

10-6 续表8 continued

沿海县 Coastal County		第一产业增加值 Value-added of Primary Industry		第二产业增加值 Value-added of Secondary Industry	
		2014	2015	2014	2015
雷 州	Leizhou	921 538	950 578	338 018	383 995
吴 川	Wuchuan	267 081	280 913	903 816	953 563
电 白	Dianbai	982 388	1 052 795	1 785 346	2 200 606
惠 东	Huidong	401 500	426 805	2 161 935	2 464 554
海 丰	Haifeng	335 164	358 032	1 129 948	1 217 964
陆 丰	Lufeng	457 228	491 261	1 029 704	1 048 885
阳 西	Yangxi	524 660	535 615	678 382	748 148
阳 东	Yangdong	415 140	436 987	1 493 432	1 586 331
饶 平	Raoping	380 244	397 902	935 549	980 189
揭 东	Jiedong	190 571	321 835	1 952 031	2 695 076
惠 来	Huilai	480 474	508 065	1 300 459	1 399 073
广 西	**Guangxi**	**878 384**	**943 151**	**886 245**	**887 990**
合 浦	Hepu	740 524	786 996	536 734	525 725
东 兴	Dongxing	137 860	156 155	349 511	362 265
海 南	**Hainan**	**5 886 753**	**5 249 151**	**3 917 689**	**3 663 217**
琼 海	Qionghai	691 510	721 525	259 124	273 935
儋 州	Danzhou	998 440		295 937	
文 昌	Wenchang	643 924	669 711	399 987	413 954
万 宁	Wanning	462 942	521 005	380 135	354 400
东 方	Dongfang	351 419	379 493	648 459	687 921
澄 迈	Chengmai	608 809	649 517	1 070 100	1 092 000
临 高	Lingao	963 935	1 016 172	87 802	91 841
昌 江	Changjiang	233 346	250 098	458 536	376 988
乐 东	Ledong	558 742	632 074	94 288	128 746
陵 水	Lingshui	373 686	409 556	223 321	243 432

10-6 续表9 continued

沿海县 Coastal County		公共财政收入 Public Revenue of Local Governments		公共财政支出 Public Expenditure of Local Governments	
		2014	2015	2014	2015
雷　州	Leizhou	55 289	58 675	353 353	553 702
吴　川	Wuchuan	60 018	68 972	234 658	316 117
电　白	Dianbai	208 432	238 197	612 036	850 149
惠　东	Huidong	302 328	347 174	528 164	654 386
海　丰	Haifeng	153 349	154 663	261 348	580 029
陆　丰	Lufeng	162 451	58 865	471 556	613 238
阳　西	Yangxi	40 425	73 496	127 241	335 012
阳　东	Yangdong	118 370	121 618	221 312	268 308
饶　平	Raoping	64 613	76 162	319 197	445 962
揭　东	Jiedong	88 792	82 088	205 999	419 962
惠　来	Huilai	60 704	83 446	303 499	456 567
广　西	**Guangxi**	**168 967**	**185 836**	**501 326**	**654 998**
合　浦	Hepu	61 701	67 349	289 876	402 512
东　兴	Dongxing	107 266	118 487	211 450	252 486
海　南	**Hainan**	**1 952 345**	**1 669 342**	**3 905 861**	**3 717 045**
琼　海	Qionghai	196 655	182 342	386 802	430 873
儋　州	Danzhou	134 023		498 342	
文　昌	Wenchang	205 560	212 983	407 732	459 660
万　宁	Wanning	142 731	164 010	385 313	414 196
东　方	Dongfang	124 138	131 999	357 712	429 060
澄　迈	Chengmai	205 115	347 607	437 687	484 346
临　高	Lingao	64 149	45 432	385 908	322 516
昌　江	Changjiang	104 234	87 297	282 228	276 059
乐　东	Ledong	124 440	103 486	335 997	423 100
陵　水	Lingshui	651 300	394 186	428 140	477 235

注:本表各省数据为合计数。

Note: The data for the provinces are the totals.

10-7 沿海地区财政收支
Financial Revenue and Expenditure by Coastal Regions

单位:亿元 （100 million yuan）

地 区 Region	地方一般公共预算收入 General Public Budget Revenue	地方一般公共预算支出 General Public Budget Expenditure
合 计 Total	**45 384.9**	**66 255.6**
天 津 Tianjin	2 667.1	3 232.4
河 北 Hebei	2 649.2	5 632.2
辽 宁 Liaoning	2 127.4	4 481.6
上 海 Shanghai	5 519.5	6 191.6
江 苏 Jiangsu	8 028.6	9 687.6
浙 江 Zhejiang	4 809.9	6 646.0
福 建 Fujian	2 544.2	4 001.6
山 东 Shandong	5 529.3	8 250.0
广 东 Guangdong	9 366.8	12 827.8
广 西 Guangxi	1 515.2	4 065.5
海 南 Hainan	627.7	1 239.4

10-8　沿海地区教育基本情况
Basic Conditions of Education by Coastal Regions

地　区 Region	高等学校数 （所） Institutions of Higher Education	本、专科在校学生数（人） Number of Students at School in Undergraduate or Specialized Courses （person）	本、专科毕（结）业生数（人） Number of Students Graduated in Undergraduate or Specialized Courses （person）
全国总计 National Total	**2 560**	**26 252 968**	**6 808 866**
天　津 Tianjin	55	512 854	132 072
河　北 Hebei	118	1 179 172	327 981
辽　宁 Liaoning	116	1 005 650	258 296
上　海 Shanghai	67	511 623	128 711
江　苏 Jiangsu	162	1 715 749	484 096
浙　江 Zhejiang	105	991 149	263 981
福　建 Fujian	88	758 452	194 652
山　东 Shandong	143	1 900 612	474 195
广　东 Guangdong	143	1 856 355	476 901
广　西 Guangxi	71	751 181	182 658
海　南 Hainan	17	182 944	48 245

10-9 沿海地区卫生基本情况

Basic Conditions of Public Health by Coastal Regions

地区 Region	卫生机构数(个) Health Institutions	医疗机构床位数(万张) Total Beds (10 000 beds)	卫生机构人员(人) Number of Employed Personnel in Health Institutions (person)
全国总计 National Total	**983 528**	**701.5**	**10 693 881**
天津 Tianjin	5 223	6.4	118 111
河北 Hebei	78 594	34.2	533 286
辽宁 Liaoning	35 236	26.7	348 525
上海 Shanghai	5 016	12.3	208 444
江苏 Jiangsu	31 925	41.4	618 945
浙江 Zhejiang	31 137	27.3	491 008
福建 Fujian	27 921	17.3	281 330
山东 Shandong	77 259	51.9	855 706
广东 Guangdong	48 320	43.6	768 482
广西 Guangxi	34 439	21.4	374 817
海南 Hainan	5 046	3.9	71 288

10-10 分地区电力消费量

Electricity Consumption by Region

单位:亿千瓦·时 (100 million kW·h)

地 区 Region	2013	2014	2015
天 津 Tianjin	774.5	794.4	800.6
河 北 Hebei	3 251.2	3 314.1	3 175.7
辽 宁 Liaoning	2 008.5	2 038.7	1 984.9
上 海 Shanghai	1 410.6	1 369.0	1 405.5
江 苏 Jiangsu	4 956.6	5 012.5	5 114.7
浙 江 Zhejiang	3 453.1	3 506.4	3 553.9
福 建 Fujian	1 700.7	1 855.8	1 851.9
山 东 Shandong	4 083.1	4 223.5	5 117.0
广 东 Guangdong	4 830.1	5 235.2	5 310.7
广 西 Guangxi	1 237.7	1 308.0	1 334.3
海 南 Hainan	232.0	251.9	272.4

10-11 沿海地区用水情况

Water Use of Coastal Regions

地　区 Region	全年供水总量 （亿立方米） Total Annual Volume of Water Supply （100 million m^3）	人均用水量 （立方米/人） Per Capita Water Use （m^3/person）
全国总计 National Total	**6 103.2**	**445.1**
天　津 Tianjin	25.7	167.8
河　北 Hebei	187.2	252.8
辽　宁 Liaoning	140.8	321.0
上　海 Shanghai	103.8	428.8
江　苏 Jiangsu	574.5	721.0
浙　江 Zhejiang	186.1	336.9
福　建 Fujian	201.3	526.6
山　东 Shandong	212.8	216.7
广　东 Guangdong	443.1	410.8
广　西 Guangxi	299.3	626.8
海　南 Hainan	45.8	504.9

10-12 沿海地区全社会固定资产投资

Total Investment in Fixed Assets by the Whole Society of Coastal Regions

单位:亿元 (100 million yuan)

地 区 Region	全社会固定资产投资 Total Investment in Fixed Assets in the Whole Country	#农、林、牧、渔业 Agriculture, Forestry, Animal Husbandry and Fishery	#交通运输、仓储和邮政业 Transport, Storage and Post
全国总计 National Total	**561 999.8**	**21 042.7**	**49 200.0**
天 津 Tianjin	11 832.0	262.6	757.2
河 北 Hebei	29 448.3	1 599.8	2 077.5
辽 宁 Liaoning	17 917.9	510.3	1 263.6
上 海 Shanghai	6 352.7	3.9	794.6
江 苏 Jiangsu	46 246.9	365.2	2 432.5
浙 江 Zhejiang	27 323.3	393.0	2 312.8
福 建 Fujian	21 301.4	638.9	2 230.7
山 东 Shandong	48 312.4	1 451.0	2 787.5
广 东 Guangdong	30 343.0	525.9	3 037.6
广 西 Guangxi	16 227.8	860.0	1 532.9
海 南 Hainan	3 451.2	58.6	421.8

10-13 沿海地区按经营单位所在地分货物进出口总额
Total Value of Imports and Exports by Location of Importers/Exporters of Coastal Regions

单位:万美元 （10 000 USD）

地　区 Region	进出口 Total	出口 Exports	进口 Imports
全国总计 National Total	**395 303 272**	**227 346 822**	**167 956 450**
天　津 Tianjin	11 428 280	5 116 293	6 311 988
河　北 Hebei	5 151 375	3 293 276	1 858 099
辽　宁 Liaoning	9 594 713	5 071 098	4 523 614
上　海 Shanghai	44 924 072	19 591 321	25 332 751
江　苏 Jiangsu	54 556 045	33 864 478	20 691 567
浙　江 Zhejiang	34 678 383	27 633 211	7 045 171
福　建 Fujian	16 884 593	11 268 011	5 616 582
山　东 Shandong	24 060 780	14 392 568	9 668 212
广　东 Guangdong	102 249 568	64 317 208	37 932 360
广　西 Guangxi	5 109 055	2 793 398	2 315 657
海　南 Hainan	1 396 697	374 304	1 022 393

10-14 沿海地区城镇居民平均每人全年家庭收入和消费性支出

Per Capita Annual Income and Consumption Expenditure of Urban Households by Coastal Regions

单位:元 (yuan)

地 区 Region	可支配收入 Disposable Income	工资性收入 Income from Wages and Salaries	消费支出 Expenses on Consumption
全国平均水平 National Average	**31 194.8**	**19 337.1**	**21 392.4**
天 津 Tianjin	34 101.3	21 060.4	26 229.5
河 北 Hebei	26 152.2	16 705.3	17 586.6
辽 宁 Liaoning	31 125.7	17 126.7	21 556.7
上 海 Shanghai	52 961.9	32 010.0	36 946.1
江 苏 Jiangsu	37 173.5	22 460.5	24 966.0
浙 江 Zhejiang	43 714.5	24 947.7	28 661.3
福 建 Fujian	33 275.3	20 714.3	23 520.2
山 东 Shandong	31 545.3	20 386.1	19 853.8
广 东 Guangdong	34 757.2	26 136.9	25 673.1
广 西 Guangxi	26 415.9	15 163.1	16 321.2
海 南 Hainan	26 356.4	17 214.9	18 448.4

10-15 沿海地区农村居民可支配收入和生活消费支出
Per Capita Disposable Income and Consumption Expenditure of Rural Residents by Coastal Regions

单位:元 (yuan)

地　区 Region	可支配收入 Disposable Income	消费支出 Consumption Expenditure
全国平均水平 National Average	**11 421.7**	**9 222.6**
天　津 Tianjin	18 481.6	14 739.4
河　北 Hebei	11 050.5	9 022.8
辽　宁 Liaoning	12 056.9	8 872.8
上　海 Shanghai	23 205.2	16 152.3
江　苏 Jiangsu	16 256.7	12 882.5
浙　江 Zhejiang	21 125.0	16 107.7
福　建 Fujian	13 792.7	11 960.8
山　东 Shandong	12 930.4	8 747.6
广　东 Guangdong	13 360.4	11 103.0
广　西 Guangxi	9 466.6	7 582.0
海　南 Hainan	10 857.6	8 210.3

10-16 沿海地区年末人口数
Year-end Population by Coastal Regions

单位:万人　　　　　　　　　　　　　　　　　　　　　　　　　　　　(10 000 persons)

地　区 Region	2013	2014	2015
全国总计 National Total	**136 072**	**136 782**	**137 462**
天　津 Tianjin	1 472	1 517	1 547
河　北 Hebei	7 333	7 384	7 425
辽　宁 Liaoning	4 390	4 391	4 382
上　海 Shanghai	2 415	2 426	2 415
江　苏 Jiangsu	7 939	7 960	7 976
浙　江 Zhejiang	5 498	5 508	5 539
福　建 Fujian	3 774	3 806	3 839
山　东 Shandong	9 733	9 789	9 847
广　东 Guangdong	10 644	10 724	10 849
广　西 Guangxi	4 719	4 754	4 796
海　南 Hainan	895	903	911

注:本表为年度人口抽样调查推算数据。2005年起各地区数据为常住人口口径。

Note: This table is estimated from the annual national sample survey on population. Since 2005, data by region are of usual residents.

10-17 沿海城市人口情况(2014年)
Population of Coastal Cities, 2014

单位:万人 (10 000 persons)

沿海城市	Coastal City	常住人口 Permanent Population	年末总人口 Total Population by the End of the Year
合　计	**Total**	**29 319.3**	**25 022.9**
天　津	**Tianjin**	**1 517.0**	**1 018.4**
河　北	**Hebei**	**1 820.8**	**1 816.7**
唐　山	Tangshan	776.8	753.2
秦皇岛	Qinhuangdao	306.5	295.1
沧　州	Cangzhou	737.5	768.4
辽　宁	**Liaoning**	**1 893.9**	**1 782.3**
大　连	Dalian	698.4	594.3
丹　东	Dandong	241.8	239.5
锦　州	Jinzhou	307.6	305.3
营　口	Yingkou	244.6	233.3
盘　锦	Panjin	144.0	129.2
葫芦岛	Huludao	257.5	280.7
上　海	**Shanghai**	**2 426.0**	**1 438.7**
江　苏	**Jiangsu**	**1 897.3**	**2 122.6**
南　通	Nantong	729.8	767.6
连云港	Lianyungang	445.2	526.5
盐　城	Yancheng	722.3	828.5

10-17 续表1 continued

沿海城市 Coastal City		常住人口 Permanent Population	年末总人口 Total Population by the End of the Year
浙 江	**Zhejiang**	**4 245.8**	**3 599.0**
杭 州	Hangzhou	889.2	715.8
宁 波	Ningbo	781.1	583.8
温 州	Wenzhou	906.8	813.7
嘉 兴	Jiaxing	457.0	348.1
绍 兴	Shaoxing	495.6	443.0
舟 山	Zhoushan	114.6	97.5
台 州	Taizhou	601.5	597.1
福 建	**Fujian**	**3 034.0**	**2 785.3**
福 州	Fuzhou	743.0	674.9
厦 门	Xiamen	381.0	203.4
莆 田	Putian	285.0	341.2
泉 州	Quanzhou	844.0	716.2
漳 州	Zhangzhou	496.0	497.4
宁 德	Ningde	285.0	352.2
山 东	**Shandong**	**3 691.4**	**3 446.8**
青 岛	Qingdao	904.6	780.6
东 营	Dongying	209.9	189.1
烟 台	Yantai	700.2	653.4
潍 坊	Weifang	924.7	888.3
威 海	Weihai	280.9	254.8
日 照	Rizhao	287.1	293.9
滨 州	Binzhou	384.0	386.7

10-17 续表2 continued

沿海城市 Coastal City		常住人口 Permanent Population	年末总人口 Total Population by the End of the Year
广 东	**Guangdong**	**7 929.5**	**6 123.7**
广 州	Guangzhou	1 308.1	842.4
深 圳	Shenzhen	1 077.9	332.2
珠 海	Zhuhai	161.4	110.2
汕 头	Shantou	552.4	546.6
江 门	Jiangmen	451.1	393.4
湛 江	Zhanjiang	721.2	819.0
茂 名	Maoming	604.9	772.4
惠 州	Huizhou	472.7	348.5
汕 尾	Shanwei	300.7	359.1
阳 江	Yangjiang	250.0	289.4
东 莞	Dongguan	834.3	191.4
中 山	Zhongshan	319.3	156.1
潮 州	Chaozhou	272.0	268.8
揭 阳	Jieyang	603.5	694.2
广 西	**Guangxi**	**569.3**	**665.5**
北 海	Beihai	160.4	169.3
防城港	Fangchenggang	90.8	94.2
钦 州	Qinzhou	318.1	402.0
海 南	**Hainan**	**294.3**	**223.9**
海 口	Haikou	220.1	165.3
三 亚	Sanya	74.2	58.6

注：本表各省数据为合计数。

Note: The data for the provinces are the totals.

10-18 沿海县人口情况
Population of Coastal Counties

单位：万人　　　　(10 000 persons)

沿海县 Coastal County		户籍总人口 Total Population of Household Registration	
		2014	2015
合　计	**Total**	**8 373**	**8 378**
河　北	**Hebei**	**331**	**339**
丰　南	Fengnan	53	53
滦　南	Luannan	57	57
乐　亭	Laoting	49	45
昌　黎	Changli	53	53
抚　宁	Funing	48	34
黄　骅	Huanghua	47	48
海　兴	Haixing	24	49
辽　宁	**Liaoning**	**658**	**657**
长　海	Changhai	7	7
瓦房店	Wafangdian	99	100
普兰店	Pulandian	92	92
庄　河	Zhuanghe	90	90
东　港	Donggang	61	61
凌　海	Linghai	52	52
盖　州	Gaizhou	71	70
大　洼	Dawa	39	38
盘　山	Panshan	28	28
绥　中	Suizhong	65	65
兴　城	Xingcheng	54	54
江　苏	**Jiangsu**	**1 303**	**1 305**
海　安	Hai′an	94	94
如　东	Rudong	104	104

10-18 续表1 continued

沿海县 Coastal County		户籍总人口 Total Population of Household Registration	
		2014	2015
启 东	Qidong	112	112
海 门	Haimen	100	100
赣 榆	Ganyu	119	120
东 海	Donghai	122	123
灌 云	Guanyun	104	105
灌 南	Guannan	81	82
响 水	Xiangshui	62	62
滨 海	Binhai	121	122
射 阳	Sheyang	97	96
东 台	Dongtai	114	113
大 丰	Dafeng	73	72
浙 江	**Zhejiang**	**1 493**	**1 496**
象 山	Xiangshan	55	55
宁 海	Ninghai	63	63
余 姚	Yuyao	84	84
慈 溪	Cixi	105	105
奉 化	Fenghua	48	48
洞 头	Dongtou	13	15
平 阳	Pingyang	88	88
苍 南	Cangnan	133	133
瑞 安	Rui´an	123	123
乐 清	Yueqing	129	128
海 盐	Haiyan	38	38
海 宁	Haining	67	68
平 湖	Pinghu	49	49
越 城	Yuecheng		
柯 桥	Keqiao	65	65

10-18 续表2 continued

沿海县 Coastal County		户籍总人口 Total Population of Household Registration	
		2014	2015
上 虞	Shangyu	78	78
岱 山	Daishan	19	19
嵊 泗	Shengsi	8	8
玉 环	Yuhuan	43	43
三 门	Sanmen	44	44
温 岭	Wenling	122	122
临 海	Linhai	119	120
福 建	**Fujian**	**1 341**	**1 356**
连 江	Lianjiang	66	67
罗 源	Luoyuan	26	27
平 潭	Pingtan	43	43
福 清	Fuqing	134	134
长 乐	Changle	71	71
仙 游	Xianyou	113	114
惠 安	Hui´an	100	101
金 门	Jinmen		
石 狮	Shishi	33	33
晋 江	Jinjiang	111	119
南 安	Nan´an	157	159
云 霄	Yunxiao	45	45
漳 浦	Zhangpu	89	90
诏 安	Zhao´an	65	65
东 山	Dongshan	21	22
龙 海	Longhai	85	86
霞 浦	Xiapu	55	54
福 安	Fu´an	67	67
福 鼎	Fuding	60	59

10-18 续表3 continued

沿海县 Coastal County		户籍总人口 Total Population of Household Registration	
		2014	2015
山　东	**Shandong**	**1 138**	**1 145**
胶　州	Jiaozhou	82	83
即　墨	Jimo	115	115
胶　南	Jiaonan		
垦　利	Kenli	23	23
利　津	Lijin	28	30
广　饶	Guangrao	51	52
长　岛	Changdao	4	4
龙　口	Longkou	64	64
莱　阳	Laiyang	86	86
莱　州	Laizhou	85	85
蓬　莱	Penglai	45	45
招　远	Zhaoyuan	57	57
海　阳	Haiyang	66	66
寿　光	Shouguang	107	107
昌　邑	Changyi	58	59
文　登	Wendeng	58	58
荣　成	Rongcheng	67	67
乳　山	Rushan	56	56
无　棣	Wudi	47	48
沾　化	Zhanhua	39	40
广　东	**Guangdong**	**1 445**	**1 518**
南　澳	Nan′ao	8	8
台　山	Taishan	98	97
恩　平	Enping	53	51
遂　溪	Suixi	109	108
徐　闻	Xuwen	76	76

10-18 续表4 continued

沿海县 Coastal County		户籍总人口 Total Population of Household Registration	
		2014	2015
廉　江	Lianjiang	179	180
雷　州	Leizhou	177	177
吴　川	Wuchuan	118	119
电　白	Dianbai	200	204
惠　东	Huidong	85	87
海　丰	Haifeng	84	85
陆　丰	Lufeng	187	187
阳　西	Yangxi	53	54
阳　东	Yangdong	50	50
饶　平	Raoping	105	107
揭　东	Jiedong	67	111
惠　来	Huilai	140	157
广　西	**Guangxi**	**120**	**122**
东　兴	Dongxing	106	107
合　浦	Hepu	14	15
海　南	**Hainan**	**544**	**440**
琼　海	Qionghai	51	51
儋　州	Danzhou	96	
文　昌	Wenchang	60	57
万　宁	Wanning	64	64
东　方	Dongfang	46	44
澄　迈	Chengmai	58	58
临　高	Lingao	51	49
昌　江	Changjiang	26	25
乐　东	Ledong	55	53
陵　水	Lingshui	37	39

注：本表各省数据为合计数。

Note: The data for the provinces are the totals.

10-19 沿海地区城镇单位就业人员情况
Employed Persons in Urban Units by Coastal Regions

单位:万人 (10 000 persons)

地 区 Region	2013	2014	2015
全国总计 National Total	**18 108.4**	**18 277.8**	**18 062.5**
天 津 Tianjin	302.4	295.5	294.8
河 北 Hebei	653.4	656.2	643.6
辽 宁 Liaoning	689.1	665.2	618.4
上 海 Shanghai	618.8	648.9	637.2
江 苏 Jiangsu	1 503.3	1 602.4	1 552.1
浙 江 Zhejiang	1 071.6	1 102.7	1 083.4
福 建 Fujian	644.0	654.6	663.1
山 东 Shandong	1 290.6	1 266.3	1 236.7
广 东 Guangdong	1 967.0	1 973.3	1 948.0
广 西 Guangxi	403.0	401.5	405.4
海 南 Hainan	98.8	101.5	100.4

10-20 沿海城市就业人员情况(2014年)

Number of Employed Persons by Coastal Cities, 2014

单位:万人 (10 000 persons)

沿海城市	Coastal City	就业人员 Number of Employed Persons	城镇就业人员 Urban Employed Persons
合　计	**Total**	**15 431.0**	**4 578.9**
天　津	**Tianjin**		
河　北	**Hebei**	**1 097.3**	**180.7**
唐　山	Tangshan	474.5	93.7
秦皇岛	Qinhuangdao	178.6	33.9
沧　州	Cangzhou	444.2	53.1
辽　宁	**Liaoning**	**1 239.8**	**284.9**
大　连	Dalian	501.2	121.3
丹　东	Dandong	129.9	28.8
锦　州	Jinzhou	178.1	32.8
营　口	Yingkou	168.2	27.2
盘　锦	Panjin	112.8	48.3
葫芦岛	Huludao	149.6	26.5
上　海	**Shanghai**		
江　苏	**Jiangsu**	**1 158.6**	**356.8**
南　通	Nantong	462.0	221.3
连云港	Lianyungang	251.1	48.2
盐　城	Yancheng	445.5	87.3
浙　江	**Zhejiang**	**2 292.4**	**942.2**
杭　州	Hangzhou	654.9	293.4
宁　波	Ningbo	511.5	171.7
温　州	Wenzhou	332.3	104.7
嘉　兴	Jiaxing	183.0	80.1
绍　兴	Shaoxing	74.3	139.7
舟　山	Zhoushan	134.2	45.2
台　州	Taizhou	402.2	107.4
福　建	**Fujian**	**2 119.8**	**571.9**
福　州	Fuzhou	503.1	149.2
厦　门	Xiamen	270.1	133.9

10-20 续表 continued

沿海城市 Coastal City		就业人员 Number of Employed Persons	城镇就业人员 Urban Employed Persons
莆 田	Putian	219.2	49.1
泉 州	Quanzhou	607.8	157.1
漳 州	Zhangzhou	314.7	52.9
宁 德	Ningde	204.9	29.7
山 东	**Shandong**	**2 458.7**	**532.9**
青 岛	Qingdao	584.7	150.1
东 营	Dongying	149.8	47.8
烟 台	Yantai	459.5	109.7
潍 坊	Weifang	569.3	86.2
威 海	Weihai	199.2	56.4
日 照	Rizhao	208.7	31.1
滨 州	Binzhou	287.5	51.6
广 东	**Guangdong**	**4 699.1**	**1 601.8**
广 州	Guangzhou	784.8	326.4
深 圳	Shenzhen	899.7	458.5
珠 海	Zhuhai	108.8	75.3
汕 头	Shantou	238.3	55.0
江 门	Jiangmen	243.2	59.9
湛 江	Zhanjiang	340.8	51.7
茂 名	Maoming	281.0	44.9
惠 州	Huizhou	280.6	91.8
汕 尾	Shanwei	119.4	24.0
阳 江	Yangjiang	128.3	24.5
东 莞	Dongguan	660.5	238.7
中 山	Zhongshan	211.8	89.5
潮 州	Chaozhou	127.7	20.6
揭 阳	Jieyang	274.2	41.0
广 西	**Guangxi**	**155.6**	**44.4**
北 海	Beihai	60.3	14.4
防城港	Fangchenggang	63.7	9.7
钦 州	Qinzhou	31.6	20.3
海 南	**Hainan**	**209.7**	**63.3**
海 口	Haikou	161.2	51.3
三 亚	Sanya	48.5	12.0

注:本表各省数据为合计数。

Note: The data for the provinces are the totals.

主要统计指标解释

1. 国内(或地区)生产总值 指一个国家(或地区)所有常驻单位在一定时期内生产活动的最终成果。国内生产总值有三种表现形态,即价值形态、收入形态和产品形态。从价值形态看,它是所有常驻单位在一定时期内生产的全部货物和服务价值超过同期中间投入的全部非固定资产货物和服务价值的差额,即所有常驻单位的增加值之和;从收入形态看,它是所有常驻单位在一定时期内创造并分配给常驻单位和非常驻单位的初次收入分配之和;从产品形态看,它是所有常驻单位在一定时期内最终使用的货物和服务价值与货物和服务净出口价值之和。在实际核算中,国内生产总值有三种计算方法,即生产法、收入法和支出法。三种方法分别从不同的方面反映国内生产总值及其构成。

2. 三次产业 是根据社会生产活动历史发展的顺序对产业结构的划分,产品直接取自自然界的部门称为第一产业,对初级产品进行再加工的部门称为第二产业,为生产和消费提供各种服务的部门称为第三产业。它是世界上较为通用的产业结构分类,但各国的划分不尽一致。我国的三次产业划分如下。

第一产业:是指农、林、牧、渔业。

第二产业:是指采矿业,制造业,电力、燃气及水的生产和供应业,建筑业。

第三产业:是指除第一、第二产业以外的其他行业。第三产业包括:交通运输、仓储和邮政业,信息传输、计算机服务和软件业,批发和零售业,住宿和餐饮业,金融业,房地产业,租赁和商务服务业,科学研究、技术服务和地质勘查业,水利、环境和公共设施管理业,居民服务和其他服务业,教育,卫生、社会保障和社会福利业,文化、体育和娱乐业,公共管理和社会组织,国际组织。

3. 增加值 是指各行各业生产经营和劳务活动的最终成果,采用生产法和收入法两种方法计算。

生产法 是从货物和服务活动在生产过程中形成的总产品入手,剔除生产过程中投入的中间产品价值,得到新增价值的方法。

收入法 又称分配法。按收入法计算国内生产总值是从生产过程创造的收入的角度对常驻单位的生产活动成果进行核算;按照此法计算,增加值由劳动者报酬、固定资产折旧、生产税净额和营业盈余四个部分组成。

4. 年末总人口 是指每年12月31日24时一定地区范围内的有生命的个人的人口总和。

5. 财政收入 指国家财政参与社会产品分配所取得的收入,是实现国家职能的财力保证。财政收入所包括的内容几经变化,目前主要包括:

(1)各项税收 包括增值税、营业税、消费税、土地增值税、城市维护建设税、资源税、城市土地使用税、企业所得税、个人所得税、关税、证券交易印花税、车辆购置税、农牧业税和耕地占用税等;

(2)专项收入 包括排污费收入、城市水资源费收入、矿产资源补偿费收入、教育费附加收入等;

(3)其他收入　包括利息收入、基本建设贷款归还收入、基本建设收入、捐赠收入等；

(4)国有企业亏损补贴　此项为负收入，冲减财政收入。主要包括对工业企业、商业企业、粮食企业的补贴。

6. 财政支出　国家财政将筹集起来的资金进行分配使用，以满足经济建设和各项事业的需要。

7. 基本建设支出　指按国家有关规定，属于基本建设范围内的基本建设有偿使用、拨款、资本金支出以及经国家批准对专项和政策性基建投资贷款，在部门的基建投资额中统筹支付的贴息支出。

8. 普通高等学校　指按照国家规定的设置标准和审批程序批准举办的，通过全国普通高等学校统一招生考试，招收高中毕业生为主要培养对象，实施高等教育的全日制大学、独立设置的学院和高等专科学校、高等职业学校和其他机构。

大学、独立设置的学院主要实施本科层次以上教育，高等专科学校、高等职业学校实施专科层次教育，其他机构是承担国家普通招生计划任务不计校数的机构。包括普通高等学校分校和批准筹建的普通高等学校等。

9. 卫生机构　包括医疗机构、疾病预防控制中心(防疫站)、采供血机构、卫生监督及监测(检验)机构、医学科研和在职培训机构、健康教育所等。

10. 医疗机构　包括医院、社区卫生服务中心(站)、疗养院、卫生院、门诊部、诊所(卫生所、医务室)、妇幼保健院(所、站)、专科疾病防治院(所、站)、急救中心(站)和临床检验中心。医疗机构分为非营利性医疗机构和营利性医疗机构。

11. 用水总量　指分配给各类用户的包括输水损失在内的毛用水量之和，不包括海水直接利用量。

12. 全社会固定资产投资　是以货币形式表现的在一定时期内全社会建造和购置固定资产的工作量以及与此有关的费用的总称。该指标是反映固定资产投资规模、结构和发展速度的综合性指标，又是观察工程进度和考核投资效果的重要依据。全社会固定资产投资按登记注册类型可分为国有、集体、个体、联营、股份制、外商、港澳台商、其他等。

13. 进出口总额　指实际进出我国国境的货物总金额。包括对外贸易实际进出口货物，来料加工装配进出口货物，国家间、联合国及国际组织无偿援助物资和赠送品，华侨、港澳台同胞和外籍华人捐赠品，租赁期满归承租人所有的租赁货物，进料加工进出口货物，边境地方贸易及边境地区小额贸易进出口货物(边民互市贸易除外)，中外合资企业、中外合作经营企业、外商独资经营企业进出口货物和公用物品，到、离岸价格在规定限额以上的进出口货样和广告品(无商业价值、无使用价值和免费提供出口的除外)，从保税仓库提取在中国境内销售的进口货物，以及其他进出口货物。该指标可以观察一个国家在对外贸易方面的总规模。我国规定出口货物按离岸价格统计，进口货物按到岸价格统计。

14. 商品经营单位所在地进、出口额　指在所在地海关注册登记的有进出口经营权的企业实际进、出口额。

15. 城镇家庭可支配收入　指家庭成员得到可用于最终消费支出和其他非义务性支出以及储蓄的总和，即居民家庭可以用来自由支配的收入。它是家庭总收入扣除交纳的所得

税、个人交纳的社会保障支出以及记账补贴后的收入。计算公式为

可支配收入=家庭总收入-交纳所得税-个人交纳的社会保障支出-记账补贴

16. 城镇家庭消费性支出　指家庭用于日常生活的支出，包括食品、衣着、家庭设备用品及服务、医疗保健、交通和通信、娱乐教育文化服务、居住、杂项商品和服务等八大类支出。

17. 总收入　指调查期内农村住户和住户成员从各种来源渠道得到的收入总和。按收入的性质划分为工资性收入、家庭经营收入、财产性收入和转移性收入。

18. 纯收入　指农村住户当年从各个来源得到的总收入相应地扣除所发生的费用后的收入总和。计算方法

纯收入=总收入-税费支出-家庭经营费用支出-生产性固定资产折旧-赠送农村亲友支出

纯收入主要用于再生产投入和当年生活消费支出，也可用于储蓄和各种非义务性支出。“农民人均纯收入”按人口平均的纯收入水平，反映的是一个地区或一个农户农村居民的平均收入水平。

19. 人口数　指一定时点、一定地区范围内有生命的个人总和。

年度统计的年末人口数指每年12月31日24时的人口数。年度统计的全国人口总数内未包括香港、澳门特别行政区和台湾省以及海外华侨人数。

20. 城镇人口　城镇人口是指居住在城镇范围内的全部常住人口

21. 就业人员　指在16周岁及以上，从事一定社会劳动并取得劳动报酬或经营收入的人员。这一指标反映了一定时期内全部劳动力资源的实际利用情况，是研究我国基本国情国力的重要指标。

Explanatory Notes on Main Statistical Indicators

1. Gross Domestic Product （**GDP**）refers to the final result of the primary distribution of the income created by all the resident units of a country （or a region）during a certain period of time. Gross domestic product is expressed in three different forms, i.e. value, income, and products respectively. The form of value refers to the total value of all products and services produced by all resident units during a certain period of time minus total value of intermediate input of materials and services of the nature of non-fixed assets or the summation of the value added of all resident units: the form of income includes all the income created by all resident units and distributed primarily to all resident and non-resident units; the form of products refers to the summation of the value of the products and services finally used and the net export value of products and services by all resident units during a given period of time. In the practice of national accounting, gross domestic product is calculated with three approaches, i.e. production approach, income approach, and expenditure approach, which reflect the gross domestic product and its composition from different aspects.

2. Three Industries Industrial structure is classified according to the sequence of historical development of social productive activities. Primary industry refers to the extraction of natural resources; secondary industry involves processing of primary products; and tertiary industry provides services of various kinds for production and consumption. The above classification is universal in the world although it varies to some extent from country to country. The three industries in China are divided as follows.

Primary industry: refers to farming, forestry, sideline production and fishery.

Secondary industry: refers to such industries as mining, manufacturing, production and supply of electric power, fuel gas and water, and construction.

Tertiary industry: refers to all the other industries not included in the primary or secondary industries. It includes such industries as communications and transportation, storage and postal service; information transmission, computer service and software; wholesale and retailing; accommodation and catering; financial service; real estate; charter business and commercial affairs service; scientific research, technological service and geological survey; water conservancy, environmental and other public facilities management; residents service and other service trades; education, health, social security and social welfare; culture, sports and entertainment business as well as public administration and social organizations, and international organizations.

3. Added Value refers to the final result of production operation and labor activities of all trades and professions, which is calculated by using the methods of production and income.

Production Method: refers to the method whereby to get the newly added value by proceeding from the gross product of goods and service activities occurring in the course of production and then rejecting the value of intermediate product input in the course of production.

Income Method: is also called the distribution method. The calculation of the gross domestic product (GDP) by the income method is the accounting of the result of productive activities of permanent units from the angle of the income created in the course of production. According to this method, the added value is composed of the payment for laborers, depreciation for fixed assets, net tax on production and business surplus.

4. Total Population by the End of the Year refers to the sum of living individuals within a particular range of area at 24:00 on December 31 of each year.

5. Government Revenue refers to the income obtained by the government finance through participating in the distribution of social products. It is the financial guarantee to ensure government functioning. The contents of government revenue have changed several times. Now it includes the following main items:

(1) Various tax revenues, including value added tax, business tax, consumption tax, land value-added tax, tax on city maintenance and construction, resources tax, tax on use of urban

land, enterprise income tax, personal income tax, tariff, stamp tax on security transactions, tax on purchase of motor vehicles, tax on agriculture and animal husbandry and tax on occupancy of cultivated land, etc.

(2) Special revenues, including revenues from the fee on sewage treatment, fee on urban water resources, fee for the compensation of mineral resources and extra-charges for education, etc.

(3) Other revenues, including revenues from interest, repayment of capital construction loan, capital construction projects, and donations and grants.

(4) Subsidies for the losses of State-owned enterprises. This is an item of negative revenue, counteracting revenues and consisting of subsidies to industrial, commercial and grain purchasing and supply enterprises.

6. Government Expenditure refers to the distribution and use of the funds which the government finance has raised, so as to meet the needs of economic construction and various causes.

7. Expenditure for Capital Construction It refers to the non-gratuitous use of, appropriation of funds for and capital outlay on capital construction in the area of capital construction. It also covers the loans on capital construction approved by the government for special purposes or policy purposes and the expenditure with discount paid in an overall way within the amount of the funds appropriated to the departments for capital construction.

8. Regular Institutions of Higher Learning refer to educational establishments set up according to the government evaluation and approval procedures, enrolling graduates from senior secondary schools and providing higher education courses and training for senior professionals. They include full-time universities, colleges, institutions of higher professional education, institutions of higher vocational education and others.

Universities and colleges primarily provide undergraduate courses; institutions of higher professional education and institutions of higher vocational education primarily provide professional trainings; and others refer to educational establishments, which are responsible for enrolling higher education students under the State Plan but not enumerated in the total number of schools, including: branch schools of universities and colleges, and universities and colleges that have been approved and under plan for construction.

9. Health Care Institutions include: medical institutions, disease prevention and control centres (epidemic prevention stations), blood gathering and supplying institutions, health supervision and inspection (check up) institutions, medicinal scientific research and on-job training institutions, health education centres and so on.

10. Medical Organizations include: hospitals, health service centres (stations) in communities, sanatoria, health centres, out-patient clinics, clinics (health stations and infirmaries), maternity and child care agencies (centres and stations), special disease prevention and curing agencies (centres and stations), first aid centres (stations) and clinical inspection centres. Medical organizations are grouped by two types: profit -making and non -profit -making medical organizations.

11. Gross Amount of Water Used refers to gross water use distributed to users, including loss during transportation, broken down into use by agriculture, industry, living consumption and ecological protection.

12. Total Investment in Fixed Assets in the Whole Country refers to the volume of activities in construction and purchases of fixed assets of the whole country and related fees, expressed in monetary terms during the reference period. It is a comprehensive indicator which shows the size, structure and growth of the investment in fixed assets, providing a basis for observing the progress of construction projects and evaluating results of investment. Total investment in fixed assets in the whole country includes, by type of ownership, the investment by State -owned units, collective -owned units, individuals, joint ownership units, share -holding units, as well as investments by entrepreneurs from foreign countries and from Hong Kong, Macao and Taiwan, and by other units.

13. Total Imports and Exports at Customs refer to the real value of commodities imported and exported across the border of China. They include the actual imports and exports through foreign trade, imported and exported goods under the processing and assembling trades and materials, supplies and gifts as aid given gratis between governments and by the United Nations and other international organizations, and contributions donated by overseas Chinese compatriots in Hong Kong and Macao and Chinese with foreign citizenship, leasing commodities owned by tenant at the expiration of leasing period, the imported and exported commodities processed with imported materials, commodities trading in border areas (excluding mutual exchange goods), the imported and exported commodities and articles for public use of the Sino -foreign joint ventures, cooperative enterprises and ventures with sole foreign investment. Also included is the import or export of samples and advertising goods for which the CIF or FOB value is beyond the permitted ceiling (excluding goods of no trading or use value and free commodities for export), imported goods sold in China from bonded warehouses and other imported or exported goods. The indicator of the total imports and exports at customs can be used to observe the total size of external trade in a country. In accordance with the stipulation of the Chinese government, imports are calculated at CIF, while exports are calculated at FOB.

14. Import-Export Value by Location of China's Foreign Trade Managing Units refers to actual value of imports and exports carried out by corporations which have been registered by the local customs house and are vested with right to run import export business.

15. Disposable Income of Urban Households refers to the actual income at the disposal of members of the households which can be used for final consumption, other non -compulsory expenditure and savings. This equals to total income minus income tax, personal contribution to social security and subsidy for keeping diaries in being a sample household. The following formula is used:

Disposable income = total household income – income tax – personal contribution to social security–subsidy for keeping diaries for a sampled household

16. Consumption Expenditure of Urban Households refers to total expenditure of households for consumption in daily life, including expenditure on the eight categories of food; clothing; household appliances and services; health care and medical services; transport and communications; recreation, education and cultural services; housing; and miscellaneous goods and services.

17. Total Income refers to the sum of income earned from various sources by the rural households and their members during the reference period, and is classified as income from wages and salaries, household operations, properties and transfers.

18. Net Income refers to the total income of rural households from all sources minus all corresponding expenses. The formula for calculation is as follows:

Net income = total income – taxes and fees paid – household operation expenses – taxes and fees – depreciation of fixed assets for production – gifts to non–rural relatives

Net income is mainly used as input for reinvestment in production and as consumption expenditure of the year, and also used for savings and non–compulsory expenses of various forms. "Per capita net income of farmers" is the level of net income averaged by population, reflecting the average income level of rural households in a given area.

19. Total Population refers to the total number of people alive at a certain point of time within a given area.

The annual statistics on total population is taken at midnight, the 31st of December, not including residents in Taiwan province, Hong Kong and Macao and overseas Chinese.

20. Urban Population refers to all people residing in cities and towns.

21. Employed Persons refers to persons aged 16 and over who are engaged in gainful employment and thus receive remuneration payment or earn business income. This indicator reflects the actual utilization of total labour force during a certain period of time and is often used for the research on China's economic situation and national power.

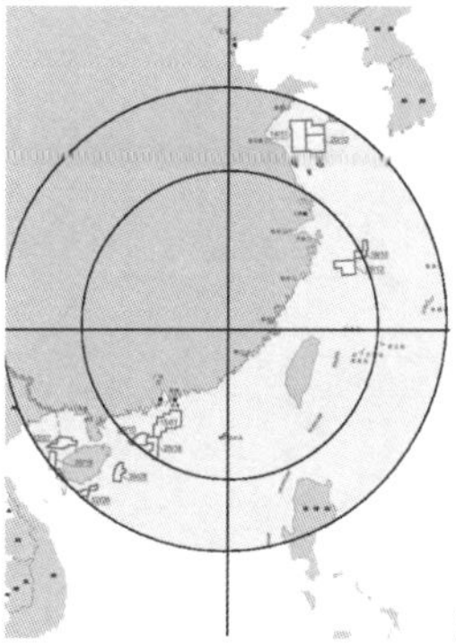

11

世界海洋经济统计资料(部分)

World's Marine Economic Statistics Data (Part)

11-1　世界海洋面积(2010年)
World Ocean Area, 2010

区 域 Region	海洋面积 (平方千米) Ocean Area (km²)	占世界海洋面积的比重(%) Proportion in the World Ocean Area (%)	占地球表面面积的比重(%) Proportion in the Earth's Surface Area (%)
合 计 **Total**	**361 000 000**	**100.0**	**70.8**
太 平 洋 Pacific Ocean	178 334 000	49.4	35.0
大 西 洋 Atlantic Ocean	91 694 000	25.4	18.0
印 度 洋 Indian Ocean	76 171 000	21.1	14.9
北 冰 洋 Arctic Ocean	14 801 000	4.1	2.9

资料来源:《2011国际统计年鉴》。

Data Source: *International Statistical Yearbook 2011.*

11-2 主要沿海国家(地区)海岸线长度
Length of Coastline of Major Coastal Countries (Areas)

单位:千米 (km)

国家或地区 Countries or Areas		海岸线长度 Length of Coastline
中国	China	32 000
美国	United States	22 680
日本	Japan	30 000
德国	Germany	1 300
英国	United Kingdom	11 450
法国	France	3 000
意大利	Italy	7 000
加拿大	Canada	20 000
澳大利亚	Australia	20 125
俄罗斯联邦	Russian Fed.	34 000
波兰	Poland	491
印度	India	6 083
印度尼西亚	Indonesia	35 000
菲律宾	Philippines	18 533
泰国	Thailand	3 219
马来西亚	Malaysia	4 675
新加坡	Singapore	193
缅甸	Myanmar	3 060
孟加拉国	Bangladesh	580
土耳其	Turkey	7 200
韩国	Korea, Rep.	2 413
埃及	Egypt	2 450
墨西哥	Mexico	9 330
巴西	Brazil	7 400
阿根廷	Argentina	4 989

11-3 世界主要沿海国家(地区)国土面积和人口
Surface Area and Population of Major Coastal Countries (Areas)

国家或地区 Countries or Areas	国土面积(万平方千米) Area of Territory (10 000 km²)	年中人口(万人) Mid-year Population (10 000 persons)	人口密度(人/平方千米) Population Density (people per km²)
世界总计 World Total	**13 432.5**	**734 663**	**57**
中国 China	960.0	137 122	146
文莱 Brunei Darussalam	0.6	42	80
柬埔寨 Cambodia	18.1	1 558	88
印度 India	298.0	131 105	441
印度尼西亚 Indonesia	191.1	25 756	142
日本 Japan	37.8	12 696	348
韩国 Korea, Rep.	10.0	5 062	519
马来西亚 Malaysia	33.1	3 033	92
缅甸 Myanmar	67.7	5 390	83
菲律宾 Philippines	30.0	10 070	338
新加坡 Singapore	0.1	554	7 829
泰国 Thailand	51.3	6 796	133
越南 Viet Nam	33.1	9 170	296
埃及 Egypt	100.2	9 151	92
南非 South Africa	121.9	5 496	45
加拿大 Canada	998.5	3 585	4
墨西哥 Mexico	196.4	12 702	65
美国 United States	983.2	32 142	35
阿根廷 Argentina	278.0	4 342	16
巴西 Brazil	851.6	20 785	25
法国 France	54.9	6 681	122
德国 Germany	35.7	8 141	234
意大利 Italy	30.1	6 080	207
俄罗斯联邦 Russian Fed.	1 709.8	14 410	9
西班牙 Spain	50.6	4 642	93
土耳其 Turkey	78.4	7 867	102
乌克兰 Ukraine	60.4	4 520	78
英国 United Kingdom	24.4	6 514	269
澳大利亚 Australia	774.1	2 378	3
新西兰 New Zealand	26.8	460	17

资料来源:《2016中国统计年鉴》。

Data Source: *China Statistical Yearbook 2016.*

11-4　主要沿海国家(地区)国内生产总值
Gross Domestic Product of Major Coastal Countries (Areas)

国家或地区 Countries or Areas	国内生产总值 (亿美元) GDP (100 million USD)	人均国内生产总值 (美元) GNI per Captia (current USD)	国内生产总值增长率 (%) Growth Rate of GDP (%)
中国　China	108 664	7 925	6.9
中国香港　Hong Kong, China	3 099	42 423	2.4
文莱　Brunei Darussalam	155	36 608	-0.5
柬埔寨　Cambodia	181	1 159	7.0
印度　India	20 735	1 582	7.6
印度尼西亚　Indonesia	8 619	3 346	4.8
日本　Japan	41 233	32 477	0.5
韩国　Korea, Rep.	13 779	27 222	2.6
马来西亚　Malaysia	2 962	9 766	5.0
菲律宾　Philippines	2 920	2 899	5.8
新加坡　Singapore	2 927	52 889	2.0
泰国　Thailand	3 953	5 816	2.8
越南　Viet Nam	1 936	2 111	6.7
埃及　Egypt	3 308	3 615	4.2
南非　South Africa	3 128	5 692	1.3
加拿大　Canada	15 505	43 249	1.1
墨西哥　Mexico	11 443	9 009	2.6
美国　United States	179 470	55 837	2.4
阿根廷　Argentina	5 481①		
巴西　Brazil	17 747	8 539	-3.9
法国　France	24 217	36 248	1.2
德国　Germany	33 558	41 219	1.7
意大利　Italy	18 148	29 847	0.8
荷兰　Netherlands	7 525	44 433	2.0
波兰　Poland	4 748	12 494	3.7
俄罗斯联邦　Russian Fed.	13 260	9 057	-3.7
西班牙　Spain	11 991	25 832	3.2
土耳其　Turkey	7 182	9 130	4.0
乌克兰　Ukraine	906	2 115	-9.9
英国　United Kingdom	28 488	43 734	2.3
澳大利亚　Australia	13 395	56 328	2.3

注:①2014年数据。

资料来源:世界银行WDI数据库。

Note: ①*Data for 2014.*

Data Source: *World Bank WDI Database.*

11-5 主要沿海国家(地区)国内生产总值产业构成
Industrial Composition of GDP of Major Coastal Countries (Areas) by Industry

国家或地区 Countries or Areas	国内生产总值产业构成(%) Industrial Structure of GDP(%)		
	第一产业 Primary Industry	第二产业 Secondary Industry	第三产业 Tertiary Industry
世界 World	**3.9**①	**27.6**①	**68.5**①
中国 China	9.0	40.5	50.5
中国香港 Hong Kong,China	0.1①	7.3①	92.7①
文莱 Brunei Darussalam	0.7②	68.2②	31.0②
柬埔寨 Cambodia	28.2	29.4	42.3
印度 India	17.4①	30.0①	52.6①
印度尼西亚 Indonesia	13.5	40.0	43.3
日本 Japan	1.2①	26.9①	72.0①
韩国 Korea, Rep.	2.3	38.0	59.7
马来西亚 Malaysia	8.4	39.1	44.3
缅甸 Myanmar			
菲律宾 Philippines	10.3	30.9	58.8
新加坡 Singapore	0.0①	24.9①	75.0①
斯里兰卡 Sri Lanka	8.7	30.7	60.6
泰国 Thailand	10.5①	36.9①	52.7①
越南 Viet Nam	17.0	33.3	39.7
埃及 Egypt	11.2	36.3	52.5
南非 South Africa	2.3	28.7	68.9
加拿大 Canada	1.8③	28.9③	69.4③
墨西哥 Mexico	3.7	32.7	63.6
美国 United States	1.3①	20.7①	78.0①
阿根廷 Argentina	6.0	27.8	66.2
巴西 Brazil	5.2	22.7	72.0
法国 France	1.7	19.5	78.8
德国 Germany	0.5	30.4	69.0
意大利 Italy	2.3	23.7	74.0
荷兰 Netherlands	1.7	20.6	77.8
俄罗斯联邦 Russian Fed.	4.6	32.6	62.8
西班牙 Spain	2.5	22.6	74.9
土耳其 Turkey	8.6	26.5	64.9
英国 United Kingdom	0.7	20.2	79.2
澳大利亚 Australia	2.4	27.1	70.5
新西兰 New Zealand	6.1③	23.0③	70.9③

注:①2014年数据。②2013年数据。③2012年数据。

资料来源:世界银行WDI数据库。

Note: ①Data for 2014. ②Data for 2013. ③Data for 2012.

Data Source: *World Bank WDI Database*.

11-6 主要沿海国家(地区)就业人数
Employment in the Major Coastal Countries (Areas)

单位:万人　　(10 000 persons)

国家或地区 Countries or Areas	2000	2005	2010	2012	2013	2014
中国 China	72 085	74 647	76 105	76 704	76 977	77 253
中国香港 Hong Kong, China	321	334	347	366	373	375
印度 India	33 020	37 199	37 429	39 267		
印度尼西亚 Indonesia	8 984	9 536	10 959	11 301	11 276	11 463
以色列 Israel	222	249	294	336	345	356
日本 Japan	6 446	6 356	6 298	6 270	6 311	6 351
韩国 Korea, Rep.	2 116	2 286	2 383	2 468	2 507	2 560
马来西亚 Malaysia	932	1 005	1 178	1 272	1 321	1 353
菲律宾 Philippines	2 745	3 231	3 604	3 760	3 792	3 865
新加坡 Singapore	209	165	306	204	206	210
斯里兰卡 Sri Lanka	631	752	771	813	842	842
泰国 Thailand	3 300	3 630	3 804	3 958	3 911	3 842
越南 Viet Nam	3 837			5 167	5 221	5 274
埃及 Egypt	1 720	1 934	2 383	2 360	2 397	2 430
南非 South Africa	1 224	1 230	1 379	1 352	1 487	1 532
加拿大 Canada	1 476	1 612	1 696	1 744	1 769	1 780
墨西哥 Mexico	3 373	4 079	4 612	4 900	4 928	4 737
美国 United States	13 689	14 173	13 906	14 247	14 393	14 631
阿根廷 Argentina	826	964	1 516	1 570	1 094	1 569
巴西 Brazil	6 563	1 955	2 202	9 471	9 666	9 945
委内瑞拉 Venezuela	896	1 073	1 207	1 257	1 295	1 319
法国 France	2 312	2 498	2 573	2 580	2 578	2 640
德国 Germany	3 632	3 636	3 799	3 913	3 953	3 987
意大利 Italy	2 093	2 241	2 253	2 257	2 219	2 228
荷兰 Netherlands	786	811	837	835	829	824
波兰 Poland	1 452	1 412	1 547	1 559	1 557	1 586
俄罗斯联邦 Russian Fed.	6 507	6 817	6 980	7 155	7 139	7 154
西班牙 Spain	1 544	1 921	1 872	1 763	1 714	1 734
土耳其 Turkey	2 158	2 205	2 259	2 482	2 552	2 593
乌克兰 Ukraine	2 018	2 068	2 027	2 035	2 040	1 807
英国 United Kingdom	2 726	2 874	2 912	2 960	2 995	3 067
澳大利亚 Australia	886	985	1 099	1 132	1 143	1 154
新西兰 New Zealand	178	208	218	222	226	231

资料来源:联合国ILO数据库。

Data Source: *ILO Database*.

11-7 主要沿海国家(地区)鱼类产量(2014年)

Fish Yields of Major Coastal Countries (Areas), 2014

单位:万吨 (10 000 tons)

国家或地区 Countries or Areas	鱼类产量 Output of Total Fishes	海域鱼类产量 Ocean Area
中国 China	3 912.4	1 142.1
印度 India	857.7	293.5
印度尼西亚 Indonesia	921.5	596.6
缅甸 Myanmar	492.9	264.6
俄罗斯联邦 Russian Fed.	416.7	380.8
美国 United States	422.4	403.0
越南 Viet Nam	483.2	218.3
日本 Japan	316.2	311.1
孟加拉国 Bangladesh	332.4	64.1
菲律宾 Philippines	282.7	238.3
泰国 Thailand	189.9	128.9
马来西亚 Malaysia	141.0	129.8
巴西 Brazil	116.7	46.2
墨西哥 Mexico	127.4	107.0
埃及 Egypt	144.7	8.8
韩国 Korea, Rep.	124.2	121.8
西班牙 Spain	107.5	105.5
尼日利亚 Nigeria	103.2	36.5
南非 South Africa	58.9	58.6
英国 United Kingdom	79.0	77.6
柬埔寨 Cambodia	71.6	9.6
土耳其 Turkey	49.9	35.7
阿根廷 Argentina	50.1	48.2
加拿大 Canada	50.7	47.1
斯里兰卡 Sri Lanka	52.2	41.7
新西兰 New Zealand	42.9	42.7
法国 France	49.2	44.7
荷兰 Netherlands	34.7	34.2

资料来源:联合国FAO数据库。

Data Source: *FAO Database*.

11-8 国家保护区面积和鱼类濒危物种

Area of National Nature Reserves and Endangered Species of Fish

国家或地区 Countries or Areas	国家保护区① National Protected①		鱼类濒危物种(种) Endangered Species of Fish (number)
	陆地保护区面积占陆地面积比重 Terrestrial Protected Areas (% of Total Land Area)	海洋保护区面积占领海面积比重 Marine Protected Areas (% of Territorial Waters)	
中国 China	17.0	2.3	131
中国香港 Hong Kong, China	41.8		13
孟加拉国 Bangladesh	4.6	2.5	21
文莱 Brunei Darussalam	44.1	1.5	8
柬埔寨 Cambodia	26.0	0.5	44
印度 India	5.4	2.1	216
印度尼西亚 Indonesia	14.7	5.8	150
伊朗 Iran	7.3	2.2	39
日本 Japan	19.4	5.1	74
韩国 Korea, Rep.	7.6	4.3	24
马来西亚 Malaysia	18.4	2.3	75
菲律宾 Philippines	11.0	2.5	77
新加坡 Singapore	5.8	1.5	26
斯里兰卡 Sri Lanka	23.2	1.3	46
泰国 Thailand	18.8	5.2	99
越南 Viet Nam	6.5	1.8	76
埃及 Egypt	11.2	13.2	47
尼日利亚 Nigeria	14.2	0.2	71
南非 South Africa	8.9	13.4	103
加拿大 Canada	9.4	1.4	42
墨西哥 Mexico	12.9	19.0	179
美国 United States	13.9	31.7	247
阿根廷 Argentina	6.8	8.9	39
巴西 Brazil	28.4	20.5	85
委内瑞拉 Venezuela	53.9	16.8	42
法国 France	25.3	62.9	52
德国 Germany	37.4	64.8	24
意大利 Italy	21.5	20.1	51
荷兰 Netherlands	11.6	57.7	15
波兰 Poland	30.0	52.7	8
俄罗斯联邦 Russian Fed.	11.4	11.5	40
西班牙 Spain	28.0	7.5	77
土耳其 Turkey	0.2	0.4	131
乌克兰 Ukraine	4.0	10.7	24
英国 United Kingdom	28.4	17.2	47
澳大利亚 Australia	14.6	48.5	111
新西兰 New Zealand	32.5	12.5	35

注:①2014年数据。

Note: ①Data refer to 2014.

资料来源:世界银行WDI数据库。

Data Source: *World Bank WDI Database*.

11-9　主要沿海国家(地区)风力发电量

Wind-Power Capacities of Major Coastal Countries (Areas)

单位:百万千瓦·时　　(million kilowatt hours)

国家或地区 Countries or Areas	2006	2007	2008	2009	2010	2011	2013
中国 China			13 079	46 096	44 622	70 331	141 197
伊朗 Iran		143	196	224			376
以色列 Israel	2	1	9	9	8		6
日本 Japan	1 753	2 624	2 623	2 949	3 962	7	5 201
韩国 Korea, Rep	239	376	436	685	817	863	1 149
马来西亚 Malaysia							
巴基斯坦 Pakistan							
菲律宾 Philippines	55	59	61	64	62	88	66
新加坡 Singapore							
斯里兰卡 Sri Lanka	2	2	3	3	53	92	236
泰国 Thailand	1				3	3	305
越南 Viet Nam						87	90
埃及 Egypt	616	831	931	1 133	1 498	1 525	1 332
尼日利亚 Nigeria							
南非 South Africa	32	32	32	32	32	103	37
加拿大 Canada	2 500	3 024	3 819	4 573	9 557	10 187	11 594
墨西哥 Mexico	59	262	269	596	1 239	1 648	4 185
美国 United States	26 676	34 603	55 696	74 226	95 148	120 854	169 713
阿根廷 Argentina	70	61	42	36	25	26	461
巴西 Brazil						2 705	6 579
委内瑞拉 Venezuela							
法国 France	2 150	4 052	5 689	7 891	9 969	12 052	16 033
德国 Germany	30 710	39 713	40 574	38 639	37 793	48 883	51 708
意大利 Italy	2 971	4 034	4 861	6 543	9 126	9 856	14 897
荷兰 Netherlands	2 733	3 438	4 260	4 581	3 993	5 100	5 627
波兰 Poland	256	522	837	1 077	1 664	3 205	6 004
俄罗斯联邦 Russian Fed.	5	7	5	4	4	5	5
西班牙 Spain	23 040	27 509	32 203	37 773	44 165	42 918	53 903
土耳其 Turkey	127	355	847	1 495	2 916	4 723	7 557
乌克兰 Ukraine	35	45	45	43	50	89	639
英国 United Kingdom	4 225	5 274	7 097	9 304	10 183	15 509	28 434
澳大利亚 Australia	1 691	2 611	3 941	3 806	4 798	5 807	7 328
新西兰 New Zealand	623	937	1 057	1 471	1 634	1 952	2 020

资料来源:联合国ESD数据库。
Data Source: UN ESD Database.

11-10 主要沿海国家(地区)捕捞产量
Fishing Yields in the Major Coastal Countries (Areas)

单位:吨 (t)

国家或地区 Countries or Areas	2013	2014
世界总计 World Total	**92 669 169**	**93 445 234**
中国 China	16 274 926	17 106 547
秘鲁 Peru	5 854 347	3 573 371
印度尼西亚 Indonesia	6 037 781	6 436 715
美国 United States	5 141 874	4 975 947
印度 India	4 645 182	4 718 821①
俄罗斯联邦 Russian Fed.	4 348 382	4 225 556
日本 Japan	3 655 650	3 660 966
缅甸 Myanmar	3 786 840	4 083 270
智利 Chile	1 770 945	2 175 486
越南 Viet Nam	2 803 800	2 919 200
菲律宾 Philippines	2 331 721	2 350 886
挪威 Norway	2 079 338	2 301 609
泰国 Thailand	1 824 829	1 769 546
韩国 Korea, Rep.	1 593 238	1 727 643
孟加拉国 Bangladesh	1 550 446	1 591 190
墨西哥 Mexico	1 615 935	1 519 893
马来西亚 Malaysia	1 488 705	1 464 071
冰岛 Iceland	1 366 675	1 076 769
西班牙 Spain	987 451	1 109 537
摩洛哥 Morocco	1 253 299	1 365 149
中国台湾 Taiwan China	925 268	1 068 278
加拿大 Canada	852 082	862 365
巴西 Brazil	765 286	767 026
阿根廷 Argentina	870 705	829 935
南非 South Africa	417 947	596 302
尼日利亚 Nigeria	721 355	759 828
英国 United Kingdom	632 315	755 690
柬埔寨 Cambodia	639 468	625 255
伊朗 Iran	559 162	627 180
丹麦 Denmark	668 480	745 146
斯里兰卡 Sri Lanka	492 103	535 050
新西兰 New Zealand	442 738	442 097
土耳其 Turkey	374 128	302 214
法国 France	496 221	521 419
埃及 Egypt	356 857	344 791
荷兰 Netherlands	327 437	369 886
委内瑞拉 Venezuela	214 161①	209 721
德国 Germany	235 107	242 743

注:捕捞品种包括鱼类、甲壳类、软体类等水生动物。①为联合国粮农组织估算值。
资料来源:《渔业和水产养殖统计年鉴》,联合国粮农组织,2014年。
Note: The fished species include fish, crustacea, mollusc and other aquatic animals.
① It is estimated by FAO from available sources of information or calculation.
Source: *Fishery and Aquaculture Statistical Yearbook*, FAO, 2014.

11-11　主要沿海国家(地区)水产养殖产量
Aquaculture Production in the Major Coastal Countries (Areas)

单位:吨　　　　(t)

国家或地区 Countries or Areas	2013	2014
世界总计 World Total	**70 260 700**	**73 783 725**
中国 China	43 549 738	45 468 960
印度 India	4 550 707	4 881 019
越南 Viet Nam	3 206 510	3 397 064
印度尼西亚 Indonesia	3 973 843	4 253 896
孟加拉国 Bangladesh	1 859 808	1 956 925
挪威 Norway	1 247 865	1 332 497
泰国 Thailand	997 515	934 758
智利 Chile	1 033 206	1 214 523
埃及 Egypt	1 097 544	1 137 091
缅甸 Myanmar	929 180	962 156
菲律宾 Philippines	815 008	788 029
巴西 Brazil	477 022	561 803
日本 Japan	608 800	657 000
韩国 Korea, Rep.	402 141	480 394
美国 United States	421 460	425 870
中国台湾 Taiwan China	344 453	339 609
伊朗 Iran	325 325	320 174
马来西亚 Malaysia	261 271	275 682
西班牙 Spain	223 708	282 238
尼日利亚 Nigeria	278 706	313 231
土耳其 Turkey	233 864	234 302
法国 France	200 410	204 000
英国 United Kingdom	203 263	204 617
加拿大 Canada	172 097	
意大利 Italy	162 620	162 550
俄罗斯联邦 Russian Fed.	154 898	161 214
墨西哥 Mexico	171 792	194 224

资料来源:《渔业和水产养殖统计年鉴》,联合国粮农组织,2014年。

Note: The cultivated varieties include fish, crustacea, mollusc and other aquatic animals.

Source: *Fishery and Aquaculture Statistical Yearbook*, FAO, 2014.

11-12 主要沿海国家(地区)石油主要指标

Main Oil Indicators of Major Coastal Countries (Areas)

国家或地区 Countries or Areas	原油探明储量(亿桶) Crude Oil Proved Reserves (100 million Barrels)	石油存量(万桶) Total Petroleum Stocks (10 000 Barrels)
世界 World	**16 628**	**349 339**
中国 China	250	
孟加拉国 Bangladesh		
文莱 Brunei Darussalam		
印度 India	57	
印度尼西亚 Indonesia	37	
伊朗 Iran	1 580	
以色列 Israel		
日本 Japan		58 000
韩国 Korea, Rep.		18 500
马来西亚 Malaysia	38	
缅甸 Myanmar		
巴基斯坦 Pakistan		
菲律宾 Philippines		
泰国 Thailand		
越南 Viet Nam	44	
埃及 Egypt	44	
尼日利亚 Nigeria	370	
南非 South Africa		
加拿大 Canada	1 730	17 000
墨西哥 Mexico	98	4 900
美国 United States	400	176 100
阿根廷 Argentina	24	
巴西 Brazil	160	
委内瑞拉 Venezuela	2 980	
法国 France		16 700
德国 Germany		29 000
意大利 Italy		12 500
荷兰 Netherlands		11 600
波兰 Poland		6 300
俄罗斯联邦 Russian Fed.	800	
西班牙 Spain		11 200
土耳其 Turkey		6 200
乌克兰 Ukraine		
英国 United Kingdom	30	7 800
澳大利亚 Australia		3 700
新西兰 New Zealand		840

注:①2013年数据。

Note: ①Data refer to 2013.

资料来源:美国能源信息署。

Data Source: U. S.Energy Information Administration.

11-13 万美元国内生产总值能耗

Energy Consumption per Ten Thousand USD of GDP

单位:吨标准油/万美元 (ton of oil equivalent per 10 000 USD)

国家或地区 Countries or Areas	2011年不变价，PPP (Constant 2011, PPP)			
	2010	2011	2012	2013
世界 World	1.37	1.34	1.32	1.31
中国 China	1.96	1.94	1.89	1.89
中国香港 Hong Kong, China	0.40	0.42	0.40	0.38
孟加拉国 Bangladesh	0.82	0.80	0.79	0.76
文莱 Brunei Darussalam	1.15	1.32	1.30	1.05
柬埔寨 Cambodia	1.47	1.43	1.39	1.34
印度 India	1.26	1.23	1.22	1.18
印度尼西亚 Indonesia	1.02	0.95	0.92	0.88
伊朗 Iran	1.60	1.57	1.74	1.85
以色列 Israel	1.03	0.97	0.99	0.95
日本 Japan	1.13	1.05	1.01	1.00
韩国 Korea, Rep.	1.66	1.67	1.65	1.61
马来西亚 Malaysia	1.26	1.24	1.20	1.29
巴基斯坦 Pakistan	1.16	1.13	1.10	1.06
菲律宾 Philippines	0.77	0.74	0.74	0.72
新加坡 Singapore	0.69	0.68	0.65	0.62
斯里兰卡 Sri Lanka	0.57	0.56	0.55	0.48
泰国 Thailand	1.30	1.29	1.29	1.33
越南 Viet Nam	1.51	1.42	1.37	1.30
埃及 Egypt	0.87	0.90	0.91	0.88
尼日利亚 Nigeria	1.47	1.48	1.50	1.42
南非 South Africa	2.31	2.25	2.17	2.13
加拿大 Canada	1.81	1.80	1.73	1.71
墨西哥 Mexico	0.97	0.97	0.96	0.96
美国 United States	1.45	1.41	1.35	1.35
巴西 Brazil	0.93	0.91	0.93	0.94
委内瑞拉 Venezuela	1.51	1.34	1.38	1.28
法国 France	1.10	1.03	1.03	1.03
德国 Germany	0.98	0.90	0.90	0.92
意大利 Italy	0.80	0.78	0.78	0.76
荷兰 Netherlands	1.10	1.00	1.03	1.02
波兰 Poland	1.23	1.18	1.12	1.11
俄罗斯联邦 Russian Fed.	2.23	2.24	2.22	2.16
西班牙 Spain	0.83	0.83	0.85	0.80
土耳其 Turkey	0.88	0.86	0.87	0.84
乌克兰 Ukraine	3.68	3.34	3.23	3.06
英国 United Kingdom	0.89	0.81	0.82	0.80
澳大利亚 Australia	1.37	1.35	1.31	1.30
新西兰 New Zealand	1.33	1.29	1.33	1.32

资料来源:世界银行WDI数据库。

Data Source: *World Bank WDI Database*.

11-14 主要沿海国家(地区)国际海运装货量和卸货量
International Maritime Freight Loaded and Unloaded in the Major Coastal Countries (Areas)

单位:万吨 (10 000 t)

国家或地区 Countries or Areas	国际海运装货量 International Ocean Shipping Loading Capacity			国际海运卸货量 International Ocean Shipping Unloading Capacity		
	2000	2005	2014	2000	2005	2014
中国香港 Hong Kong, China	6 770	8 918	10 375	10 693	14 095	15 281
孟加拉国 Bangladesh	89	79	581①	1 408		4 820①
文莱 Brunei Darussalam	10	8		102	168	
印度尼西亚 Indonesia	14 153	27 372		4 504	8 479	
伊朗 Iran	3 065	3 446		4 486	6 073	
以色列 Israel	1 387	1 663	1 867	2 920	2 107	2 952
日本 Japan	13 010			80 654		
韩国 Korea, Rep.	15 078	24 250		41 882	51 245	
马来西亚 Malaysia	5 483	7 940	13 253②	6 922	10 391	15 770②
巴基斯坦 Pakistan	617	1 126		3 080	3 955	
新加坡 Singapore	32 618	42 266				
斯里兰卡 Sri Lanka	919	1 315				
埃及 Egypt		2 123			4 241	
南非 South Africa	2 598					
美国 United States	34 334	40 534		83 352	94 160	
阿根廷 Argentina	1 550					2 742①
法国 France	6 810	10 066	10 770	20 273	22 710	18 719
德国 Germany	8 602	10 832	11 795	14 725	16 866	16 967
荷兰 Netherlands	9 940	12 250		32 507	36 424	
波兰 Poland	3 152	3 835	2 929	1 582	1 642	3 919
俄罗斯联邦 Russian Fed.	828	910		84	74	
西班牙 Spain	5 627	8 195		19 343	26 164	
乌克兰 Ukraine	4 271	7 070	10 114①	684	1 333	1 610①
澳大利亚 Australia	48 750	62 401	136 933	5 418	6 989	9 881
新西兰 New Zealand	2 214	2 167	3 614	1 379	1 844	2 118

注:①2014年数据。②2013年数据。
资料来源:联合国统计月报数据库。
Note: ①Data refer to 2014. ②Data refer to 2013.
Data Source: *UN Monthly Bulletin of Statistics Database*.

11-15 世界主要外贸货物海运量及构成(2014年)
World Major Maritime Freight Traffic in Foreign Trade, 2014

品 种 Sort	海运量(百万吨) Freight Traffic (million tons)	
	2014	所占比例(%) Percentage
合 计 Total	**10 250**	**100.0**
原 油 Crude Oil	1 809	17.6
成品油 Refined Oil	980	9.6
燃 气 Fuel gas	317	3.1
铁矿石 Ironstone	1 328	13.0
煤 炭 Coal	1 196	11.7
谷 物 Corn	414	4.0
其他货物 Others	4 206	41.0

注：本表为估计数。

资料来源：Shipping Statistics and Market Review, January/February 2015, ISL

Note: This table is estimated data.

Data Source: Shipping Statistics and Market Review, January/February 2015, ISL.

11-16 主要沿海国家(地区)国际旅游人数

Number of International Tourists of Major Coastal Countries (Areas)

单位:万人 (10 000 persons)

国家或地区 Countries or Areas	入境(过夜)旅游人数 Number of Inbound Tourists (Overnight)			出境旅游人数 Number of Outbound Tourists		
	2000	2010	2013	2000	2010	2013
世界总计 World Total	**69 471**	**96 719**	**116 091**	**81 197**	**103 203**	**122 016**
中国 China	3 123	5 566	5 562	1 047	5 739	
中国香港 Hong Kong, China	881	2 009	2 777	5 890	8 444	8 452
中国澳门 Macao, China	520	1 193	1 457	14	75	154
孟加拉国 Bangladesh	20	30	13	113	191	
文莱 Brunei Darussalam	98	21	20			
柬埔寨 Cambodia	47	251	450	4	51	96
印度 India	265	578	768	442	1 299	1 833
印度尼西亚 Indonesia	506	700	944	221	624	877
伊朗 Iran	134	294	497	229		770
以色列 Israel	242	280	293	353	427	518
日本 Japan	476	861	1 341	1 782	1 664	1 690
韩国 Korea, Rep.	532	880	1 420	551	1 249	1 608
马来西亚 Malaysia	1 022	2 458	2 744	3 053		
缅甸 Myanmar	42	79	308			
巴基斯坦 Pakistan	56	91				
菲律宾 Philippines	199	352	483	167		
新加坡 Singapore	606	916	1 186	444	734	890
斯里兰卡 Sri Lanka	40	65	153	52	112	131

11-16 续表 continued

国家或地区 Countries or Areas	入境(过夜)旅游人数 Number of Inbound Tourists (Overnight)			出境旅游人数 Number of Outbound Tourists		
	2000	2010	2013	2000	2010	2013
泰国 Thailand	958	1594	2481	191	545	644
越南 Viet Nam	214	505	787			
埃及 Egypt	512	1 405	963	296	462	618
尼日利亚 Nigeria	81	156				
南非 South Africa	587	807	955	383	517	
加拿大 Canada	1 963	1 622	1 654	1 918	2 868	3 352
墨西哥 Mexico	2 064	2 329	2 935	1 108	1 433	1 826
美国 United States	5 124	6 001	7 501	6 133	6 106	6 830
阿根廷 Argentina	291	533	593	495	531	652
巴西 Brazil	531	516	643	323	645	905
委内瑞拉 Venezuela	47	53	86	95	148	159
法国 France	7 719	7 665	8 377	1 989	2 504	2 818
德国 Germany	1 898	2 688	3 300	7 440		8 301
意大利 Italy	4 118	4 363	4 858	2 199	2 982	2 846
荷兰 Netherlands	1 000	1 088	1 393	1 390	1 837	1 793
波兰 Poland	1 740	1 247	1 600	5 668	4 276	5 600
俄罗斯联邦 Russian Fed.	2 117	2 228	3 242	1 837	3 932	4 589
西班牙 Spain	4 640	5 268	6 500	410	1 238	1 178
土耳其 Turkey	959	3 136	3 981	528	656	798
乌克兰 Ukraine	643	2 120	1 271	1 342	1 718	2 244
英国 United Kingdom	2 321	2 830	3 261	5 684	5 556	6 008
澳大利亚 Australia	493	579	687	350	710	911
新西兰 New Zealand	178	244	277	128	203	228

资料来源:世界银行WDI数据库。

Data source: *WDI database of World Bank.*

11-17 主要沿海国家(地区)国际旅游收入
International Tourism Receipts of Major Coastal Countries (Areas)

单位:亿美元　　　　(10 000 million USD)

国家或地区 Countries or Areas	国际旅游收入 International Tourism Receipts				
	2010	2011	2012	2013	2014
世界总计 World Total	**11 046**	**12 495**	**12 972**	**13 811**	**14 340**
中国 China	458	533	549	564	569
中国香港 Hong Kong, China	272	337	380	426	460
中国澳门 Macao, China	282	390	445	523	516
孟加拉国 Bangladesh	1	1	1	1	2
文莱 Brunei Darussalam					
柬埔寨 Cambodia	17	18	20	29	32
印度 India	145	177	183	190	208
印度尼西亚 Indonesia	76	90	95	103	116
伊朗 Iran	26	26			
以色列 Israel	58	60	62	65	64
日本 Japan	154	125	162	169	208
韩国 Korea, Rep.	144	175	197	193	230
马来西亚 Malaysia	182	196	203	210	226
缅甸 Myanmar	1	3		9	16
巴基斯坦 Pakistan	10	11	10	9	10
菲律宾 Philippines	34	40	49	56	61
新加坡 Singapore	142	181	193	191	192
斯里兰卡 Sri Lanka	10	14	18	25	33

11-17 续表 continued

国家或地区 Countries or Areas	国际旅游收入 International Tourism Receipts				
	2010	2011	2012	2013	2014
泰国 Thailand	238	309	377	460	421
越南 Viet Nam	45	57	68	75	73
埃及 Egypt	136	93	108	73	80
尼日利亚 Nigeria	7	7	6		6
南非 South Africa	103	107	112	105	105
加拿大 Canada	184	200	207	177	175
墨西哥 Mexico	126	125	133	143	166
美国 United States	1 680	1 845	2 001	2 148	2 208
阿根廷 Argentina	56	61	57	50	52
巴西 Brazil	60	68	69	70	74
委内瑞拉 Venezuela	8	8	9		
法国 France	562	660	635	661	668
德国 Germany	491	534	516	552	559
意大利 Italy	384	454	430	462	456
荷兰 Netherlands	117	210	205	227	147
波兰 Poland	100	116	118	125	123
俄罗斯联邦 Russian Fed.	132	170	179	202	195
西班牙 Spain	543	677	632	676	651
土耳其 Turkey	263	301	323	349	374
乌克兰 Ukraine	47	54	60	60	23
英国 United Kingdom	402	459	460	494	628
澳大利亚 Australia	311	342	341	334	341
新西兰 New Zealand	65	55	55	75	84

资料来源:世界银行WDI数据库。

Data Source: *World Bank WDI Database*.

11-18 集装箱吞吐量居世界前20位的港口
World Top 20 Seaports in Terms of the Number of Containers Handled

单位:万标准箱 (10 000 TEU)

港　口 Seaport	所属国家或地区 Countries or Areas	吞吐量 Containers Handled	备　注 Remark
上海 Shanghai	中国 China	3 654	内外贸货物 Domestic and Foreign Trade Cargo
新加坡 Singapore	新加坡 Singapore	3 092	内外贸货物 Domestic and Foreign Trade Cargo
深圳 Shenzhen	中国 China	2 420	内外贸货物 Domestic and Foreign Trade Cargo
宁波—舟山 Ningbo and Zhoushan	中国 China	2 063	内外贸货物 Domestic and Foreign Trade Cargo
香港 Hong Kong	中国 China	2 008	内外贸货物 Domestic and Foreign Trade Cargo
釜山 Pusan	韩国 Korea, Rep.	1 943	内外贸货物 Domestic and Foreign Trade Cargo
青岛 Qingdao	中国 China	1 744	内外贸货物 Domestic and Foreign Trade Cargo
广州 Guangzhou	中国 China	1 740	内外贸货物 Domestic and Foreign Trade Cargo
迪拜 Dubayy	阿联酋 United Arab Emirates	1 559	内外贸货物 Domestic and Foreign Trade Cargo
天津 Tianjin	中国 China	1 411	内外贸货物 Domestic and Foreign Trade Cargo
鹿特丹 Rotterdam	荷兰 Netherlands	1 223	内外贸货物 Domestic and Foreign Trade Cargo
巴生 Kelang	马来西亚 Malaysia	1 187	内外贸货物 Domestic and Foreign Trade Cargo
高雄 Gaoxiong	中国台湾 Taiwan, China	1 026	内外贸货物 Domestic and Foreign Trade Cargo
安特卫普 Antwerp	比利时 Belgium	965	内外贸货物 Domestic and Foreign Trade Cargo
大连 Dalian	中国 China	945	内外贸货物 Domestic and Foreign Trade Cargo
厦门 Xiamen	中国 China	918	内外贸货物 Domestic and Foreign Trade Cargo
丹戎帕拉帕斯 Tanjung Periuk	马来西亚 Malaysia	912	内外贸货物 Domestic and Foreign Trade Cargo
汉堡 Hamburg	德国 Germany	880	内外贸货物 Domestic and Foreign Trade Cargo
洛杉矶 Los Angeles	美国 United States	816	内外贸货物 Domestic and Foreign Trade Cargo
长滩 Long Beach	美国 United States	719	内外贸货物 Domestic and Foreign Trade Cargo

资料来源:上海航运交易所。

Data Source: *Shanghai Shipping Exchange*.

11-19 港口货物吞吐量居世界前20位的港口(2014年)

World Top 20 Seaports in Terms of the Cargo Handled, 2014

单位:百万吨 (million t)

港口 Seaport	所属国家或地区 Countries or Areas	吞吐量 Cargo Handled	备注 Remark
上海 Shanghai	中国 China	755.3	内外贸货物 Domestic and Foreign Trade Cargo
新加坡 Singapore	新加坡 Singapore	581.3	内外贸货物 Domestic and Foreign Trade Cargo
天津 Tianjin	中国 China	540.0	内外贸货物 Domestic and Foreign Trade Cargo
宁波 Ningbo	中国 China	526.5	内外贸货物 Domestic and Foreign Trade Cargo
广州 Guangzhou	中国 China	482.2	内外贸货物 Domestic and Foreign Trade Cargo
青岛 Qingdao	中国 China	468.0	内外贸货物 Domestic and Foreign Trade Cargo
黑德兰港 Headland Harbour	澳大利亚 Australia	446.9	内外贸货物 Domestic and Foreign Trade Cargo
鹿特丹 Rotterdam	荷兰 Netherlands	444.7	内外贸货物 Domestic and Foreign Trade Cargo
大连 Dalian	中国 China	423.4	内外贸货物 Domestic and Foreign Trade Cargo
釜山 Pusan	韩国 Korea, Rep.	335.4	内外贸货物 Domestic and Foreign Trade Cargo
香港 Hong Kong	中国 China	297.7	外贸货物 Foreign Trade Cargo
秦皇岛 Qinhuangdao	中国 China	274.0	内外贸货物 Domestic and Foreign Trade Cargo
南路易斯安娜 Southern Louisiana	美国 United States	264.8	内外贸货物 Domestic and Foreign Trade Cargo
休斯敦 Houston	美国 United States	229.2	外贸货物 Foreign Trade Cargo
深圳 Shenzhen	中国 China	223.2	内外贸货物 Domestic and Foreign Trade Cargo
巴生 Kelang	马来西亚 Malaysia	217.3	内外贸货物 Domestic and Foreign Trade Cargo
名古屋 Nagoya	日本 Japan	207.6	内外贸货物 Domestic and Foreign Trade Cargo
厦门 Xiamen	中国 China	205.0	内外贸货物 Domestic and Foreign Trade Cargo
安特卫普 Antwerp	比利时 Belgium	199.0	内外贸货物 Domestic and Foreign Trade Cargo
洛杉矶 Los Angeles	美国 United States	176.5	内外贸货物 Domestic and Foreign Trade Cargo

资料来源:Shipping Statistics and Market Review, December 2015, ISL.

Data Source: *Shipping Statistics and Market Review, December 2015, ISL.*

11-20 海上商船拥有量居世界前20位的国家或地区(2014年)

World Top 20 Countries or Areas in Terms of the Number of Maritime Merchant Ships Owned ,2014

国家或地区 Countries or Areas	艘数(艘) Number of Vessels	总吨 Gross Ton (万吨) (10 000 tons)	载重吨 Deadweight ton	
			万吨 10 000 tons	占世界% Percentage in the World
世界总计 World Total	**50 422**	**110 476.1**	**166 100.8**	**100.0**
巴拿马 Panama	6 745	22 147.2	34 295.0	20.6
利比里亚 Liberia	2 996	12 708.7	19 880.0	12.0
马绍尔群岛 Marshall Islands	2 345	10 343.2	16 858.4	10.1
中国香港 Hong Kong, China	2 271	9 227.4	15 033.2	9.1
新加坡 Singapore	2 208	7 395.7	11 231.7	6.8
马耳他 Malta	1 757	5 305.0	8 150.3	4.9
希腊 Greece	1 044	4 404.6	7 819.6	4.7
中国 China	2 818	4 616.6	7 244.3	4.4
巴哈马 Bahamas	1 132	4 925.2	6 330.8	3.8
英国 United Kingdom	797	2 776.6	3 776.4	2.3
塞浦路斯 Cyprus	829	2 081.3	3 295.4	2.0
日本 Japan	2 524	1 608.0	2 155.4	1.3
挪威 Norway	781	1 446.6	1 828.9	1.1
意大利 Italy	732	1 592.6	1 694.5	1.0
丹麦 Denmark	462	1 423.5	1 658.1	1.0
韩国 Korea, Rep.	1 023	1 011.7	1 632.7	1.0
印度 India	799	859.0	1 471.9	0.9
印度尼西亚 Indonesia	2 535	930.6	1 299.0	0.8
安提瓜和巴布亚 Antigua and Papua	1 117	978.3	1 262.4	0.8
德国 Germany	316	1 134.6	1 255.0	0.8

注:统计范围为300总吨及以上船舶。统计截止日期为2015年1月1日。

资料来源:SSMR, Jan/Feb 2015, ISL (SSMR 2014年商船数据来源于 Clarkson Research Kservices Limited, CRSL).

Note: The statistical range of 300 tons and above ships. The closing date of statistics was 1 Jan., 2015.

Data Source: *SSMR, Jan/Feb* 2015, *ISL.* (*The SSMR* 2014 *merchant ship data is derived from Clarkson Research Kservices Limited, CRSL*).

11-21 集装箱船拥有量居世界前20位的国家或地区（2014年）
World Top 20 Countries or Areas in Terms of the Number of Container Ships Owned, 2014

国家或地区 Countries or Areas	艘数 （艘） Number of Vessels	载重吨 Deadweight ton	
		占世界% Percentage in the World	万标准箱 10 000 TEU
世界总计 World Total	**5 097**	**1 820.6**	**100.0**
利比里亚 Liberia	924	360.8	19.8
巴拿马 Panama	668	288.8	15.9
中国香港 Hong Kong, China	413	206.3	11.3
新加坡 Singapore	472	186.7	10.3
马绍尔群岛 Marshall Islands	275	94.0	5.2
德国 Germany	168	91.9	5.0
丹麦 Denmark	112	87.2	4.8
马耳他 Malta	199	85.6	4.7
英国 United Kingdom	124	75.6	4.2
安提瓜和巴布亚 Antigua and Papua	349	48.1	2.6
中国 China	181	47.8	2.6
塞浦路斯 Cyprus	199	39.2	2.2
美国 United States	67	24.1	1.3
法国 France	25	18.6	1.0
葡萄牙 Portugal	59	16.8	0.9
卢森堡 Luxembourg	26	15.1	0.8
希腊 Greece	18	14.8	0.8
中国台湾 Taiwan, China	36	12.6	0.7
印度尼西亚 Indonesia	183	11.2	0.6
巴哈马 Bahamas	49	11.1	0.6

注：1. 统计截止日期为2015年1月1日。

2. 因统计口径不同，表中数据与其他出版物公布的数据略有差别。

资料来源：SSMR May/June 2015, ISL（SSMR2014年集装箱船数据来源于Clarkson Research Services Limited, CRSL）。

Note: 1. The closing date of statistics was 1 Jan., 2015.

2. Owing to different statistical requirements, the data given in the table may be somewhat different from those issued in other publications.

Data Source: *SSMR May/June* 2015, *ISL*. （The SSMR 2014 container ship data is derived from *Clarkson Research Services Limited, CRSL*）.

11-22 海上油轮拥有量居世界前20位的国家或地区(2014年)
Top 20 Countries or Areas in the World by the Possession of Ocean-Going Oil Tankers, 2014

国家或地区 Countries or Areas	艘数(艘) Number of Vessels	总吨 Gross Ton (千吨) (1 000 tons)	载重吨 Deadweight ton	
			千吨 1 000 tons	占世界% Percentage in the World
世界总计 World Total	**13 647**	**351 128**	**579 205**	**100.0**
马绍尔群岛 Marshall Islands	961	47 310	77 134	13.3
巴拿马 Panama	1 553	45 864	76 186	13.2
利比里亚 Liberia	932	43 264	75 527	13.0
希腊 Greece	451	26 625	48 683	8.4
巴哈马 Bahamas	384	25 888	40 391	7.0
新加坡 Singapore	989	22 743	38 783	6.7
中国香港 Hong Kong, China	429	17 878	31 738	5.5
马耳他 Malta	565	17 336	29 640	5.1
英国 United Kingdom	271	8 400	13 217	2.3
中国 China	721	7 822	12 966	2.2
坦桑尼亚 Tanzania	64	5 844	11 133	1.9
挪威 Norway	219	7 484	11 105	1.9
印度 India	128	4 788	8 467	1.5
意大利 Italy	254	4 422	7 268	1.3
日本 Japan	862	5 690	6 899	1.2
百慕大 Bermuda	75	6 302	6 497	1.1
塞浦路斯 Cyprus	124	3 732	6 043	1.0
马来西亚 Malaysia	191	4 630	6 040	1.0
丹麦 Denmark	165	3 515	5 520	1.0
印度尼西亚 Indonesia	533	3 297	5 262	0.9

注:统计截止日期为2015年1月1日。
资料来源:Total tanker fleet by ownership patterns, SSMR March 2015, ISL.
Note: The closing date of statistics was 1 Jan., 2015.
Data Source: *Total tanker fleet by ownership patterns, SSMR March* 2015, *ISL.*

11-23 海上散货船拥有量居世界前20位的国家或地区(2014年)

Top 20 Countries or Areas in the World by the Possession of Ocean Bulk Carriers, 2014

国家或地区 Countries or Areas	艘数(艘) Number of Vessels	总吨(千吨) (1 000 tons)	载重吨 Deadweight ton	
			千吨 1 000 tons	占世界% Percentage in the World
世界总计 World Total	**10 696**	**406 100**	**738 612**	**100.0**
巴拿马 Panama	2 639	113 827	209 341	28.3
中国香港 Hong Kong, China	1 061	48 318	88 476	12.0
马绍尔群岛 Marshall Islands	927	41 391	75 783	10.3
利比里亚 Liberia	853	39 795	72 958	9.9
中国 China	1 083	28 432	48 471	6.6
新加坡 Singapore	473	24 161	44 812	6.1
马耳他 Malta	551	20 783	37 595	5.1
希腊 Greece	268	14 466	27 293	3.7
塞浦路斯 Cyprus	288	11 011	19 981	2.7
巴哈马 Bahamas	256	8 587	15 094	2.0
英国 United Kingdom	126	7 289	13 644	1.8
日本 Japan	393	6 388	11 843	1.6
韩国 Korea, Rep.	177	6 139	11 416	1.5
意大利 Italy	79	3 508	6 437	0.9
印度 India	100	2 651	4 770	0.6
菲律宾 Philippines	91	2 535	4 292	0.6
土耳其 Turkey	89	2 410	4 175	0.6
挪威 Norway	56	2 180	3 860	0.5
比利时 Belgium	21	1 568	3 019	0.4
中国台湾 Taiwan, China	39	1 496	2 773	0.4

注:统计截止日期为2015年1月1日。

资料来源:Total bulk carrier fleet by ownership patterns, SSMR April 2015, ISL.

Note: The closing date of statistics was 1 Jan., 2015.

Data Source: *Total bulk carrier fleet by ownership patterns, SSMR April 2015, ISL.*

图书在版编目(CIP)数据

中国海洋统计年鉴. 2016 : 汉、英 / 国家海洋局编. — 北京：海洋出版社，2017. 7
ISBN 978-7-5027-9833-8

Ⅰ. ①中… Ⅱ. ①国… Ⅲ. ①海洋—统计资料—中国—2016—年鉴—汉、英 Ⅳ. ①P7-66

中国版本图书馆CIP数据核字（2017）第170229号

责任编辑：王　溪

责任印制：赵麟苏

海洋出版社 出版发行

http: //www.oceanpress.com.cn

北京市海淀区大慧寺路 8 号　　　　邮编：100081

国家海洋信息中心印刷厂印刷　　　新华书店发行所经销

2017 年 7 月第 1 版　　　2017 年 7 月第 1 次印刷

开本：787mm×1092mm　1/16　印张：19.75

字数：495 千字

定价：168.00 元

发行部：（010）62147016　　　邮购部：（010）68038093　　　总编室：（010）62114335

海洋版图书印、装错误可随时退换